Carbon Capture

Edited by Syed Abdul Rehman Khan

Published in London, United Kingdom

IntechOpen

Supporting open minds since 2005

Carbon Capture
http://dx.doi.org/10.5772/intechopen.78885
Edited by Syed Abdul Rehman Khan

Assistant to the Editor: Zhang Yu

Contributors
Catinca Secuianu, Sergiu Sima, Yıldırım İsmail İsmail Tosun, Helen Onyeaka, Abarasi Hart, Ryle S. Perera, Syed Abdul Rehman Khan, Laeeq Razzak Janjua, Zhang Yu

First published in London, United Kingdom, 2021 by IntechOpen
IntechOpen is the global imprint of INTECHOPEN LIMITED, registered in England and Wales, registration number: 11086078, 5 Princes Gate Court, London, SW7 2QJ, United Kingdom
Printed in Croatia

British Library Cataloguing-in-Publication Data
A catalogue record for this book is available from the British Library

Additional hard and PDF copies can be obtained from orders@intechopen.com

Carbon Capture
Edited by Syed Abdul Rehman Khan
p. cm.
Print ISBN 978-1-78985-725-2
Online ISBN 978-1-78985-853-2
eBook (PDF) ISBN 978-1-78985-854-9

For EU product safety concerns:
IN TECH d.o.o., Prolaz Marije Krucifikse Kozulić 3, 51000 Rijeka, Croatia, info@intechopen.com or visit our website at intechopen.com.

Meet the editor

Syed Abdul Rehman Khan is a teacher of Supply Chain and Logistics Management. Dr. Khan achieved his Certified Supply Chain Professional (CSCP) certificate from the United States and successfully completed his Ph.D. in China. Since 2018, Dr. Khan has been affiliated with Xuzhou University of Technology, China, as an associate professor. He has more than nine years of core experience in supply chain and logistics at industry and academic levels. He has attended several international conferences and has been invited as a keynote speaker in different countries. He has published more than 100 scientific research papers in well-renowned international journals and conferences. He is a regular contributor to conferences and workshops worldwide. In addition, he has received two consecutive scientific innovation awards from the Education Department of Shaanxi Provincial Government, China. Dr. Khan is also a member of the American Production and Inventory Control Society, Production and Operations Management Society, Council of Supply Chain Management Professionals, Supply Chain Association of Pakistan, and Global Supply Chain Council, China.

Contents

Preface

Unplanned industrialization is compromising environmental sustainability, creating harmful effects on fauna and flora as well as causing innumerable diseases in humans. There is no doubt that economic growth depends heavily on the transportation and manufacturing industries, both of which are major sources of carbon emissions. As such, the question arises of how to control and mitigate carbon emissions without suffering a decline in economic development.

Carbon capture and storage is a climate change mitigation technology in which CO_2 is captured from power plants and other industrial processes instead of emitted to the atmosphere. Green practices in business operations, such as ecological design, green manufacturing, green transportation systems, and renewable energy and recycling techniques, can efficiently reduce pollution and waste.

This book is unique, as other books about pollution and carbon primarily focus on the global scale and emphasize environmental problems. This book, however, discusses carbon and environmental issues at domestic levels, highlights environmental problems, and provides solutions for controlling carbon emissions. Written by distinguished researchers from prestigious universities, chapters include case studies from real-world events that challenge the brightest minds.

We would like to thank the contributors for their extraordinary work. We would also like to thank Author Service Manager Ms. Sandra Maljavac for her milestone coordination and author facilitation. We also extend our thanks to IntechOpen for all of their support.

Thank you,

Syed Abdul Rehman Khan
Tsinghua University,
China

Xuzhou University of Technology,
China

Section 1

Introduction to Carbon Capture

Chapter 1

Introductory Chapter: Carbon Capture

Syed Abdul Rehman Khan and Zhang Yu

1. Introduction

Environmental degradation is not a new problem for this world, while the intensity of the problem is increasing due to non-seriousness and lack of an appropriate plan to protect the natural resources and nourish the earth's sustainability [1, 2]. There are some questions, which need to be answered before making any plan for environmental sustainability. First of all, whether are we serious about protecting the natural resources of the world? Second, can we compromise on economic growth for the sake of environmental sustainability? After answering these two questions, governmental bodies can develop a master plan to address environmental issues effectively.

2. Are we serious?

There is no doubt that the industrial revolution is a prominent cause of global warming and climate change. However, in reality, the problem is older than the industrial revolution. Besides, the industrial revolution only increased the speed of environmental degradation.

Environmental problems cannot be solved overnight. It will take time to restore and nourish environmental beauty through creating awareness in public, which will create indirect pressure on firms to produce eco-friendly products for the sake of their consumers' demand [3–5]. There are many other pro-environmental policies, which can help to improve environmental sustainability. The critical factor, which plays a vital role, is governmental bodies' willingness and seriousness to address the problem.

3. Can we compromise?

The consumption of fossil fuel and energy is not only fueling economic growth but also environmental problems. This is a primary cause why the environmental policies failed in the execution phase, and governmental bodies are unable to enforce it. According to Munir et al. [6], Economic growth and environmental sustainability have an inverse relationship. They also highlighted that economic growth is heavily based on the cost of environmental sustainability in developing nations. Wang and Zhu [7] argue that developed countries shifted their low-tech and labor-intensive manufacturing in developing nations, which is also a significant cause of social and environmental problems. Several researchers highlighted that China is a world-factory and fulfilling the demand of the European and Western world, while China's social and environmental performance is the worst in the world [8, 9]. Similarly, India and other developing countries are blindly following China to improve their economic performance without catering to social and environmental issues facing the Chinese people and government [3, 6, 10, 11].

4. Conclusion and policy recommendations

To address environmental problems without compromising economic growth, governmental bodies need to prepare a master plan, promoting eco-environmental sustainability. Following are some recommendations:

- Technological advancement in manufacturing practices is highly recommended to reduce carbon emissions.

- Adopting green practices in logistical and supply chain operations will help mitigate the harmful effect on environmental sustainability.

- The continuous invention and innovation process renews the technological process that moves towards green development.

- Provide loans to enterprises for green projects at low-interest rates.

- Increase awareness of environmental issues in public, which will help to create demand for green products.

- The green capital financing in agriculture food production may strengthen the agriculture industry, which will reduce the pressure on the environment.

- Labor productivity may further be improved by providing eco-friendly training and awareness of environmental problems.

- The financial and trade liberalization policies must be scrutinized in a serious node, as both are considered the critical determinants of globalization; thus, the tight environmental policies will effectively reduce carbon emissions and enhance environmental sustainability. Further, governmental bodies need to encourage firms for ISO certifications and eco-friendly practices through financial benefits.

Author details

Syed Abdul Rehman Khan[1]* and Zhang Yu[2]

1 School of Management and Engineering, Xuzhou University of Technology, Xuzhou, China

2 School of Economics and Management, Chang'an University, Xi'an, China

*Address all correspondence to: sarehman_cscp@yahoo.com

References

[1] Ahmed K. Environmental policy stringency, related technological change, and emissions inventory in 20 OECD countries. Journal of Environmental Management. 2020;**274**:111209

[2] Akadiri SS, Lasisi TT, Uzuner G, Akadiri AC. Examining the causal impacts of tourism, globalization, economic growth, and carbon emissions in tourism island territories: Bootstrap panel granger causality analysis. Current Issues in Tourism. 2020;**23**(4):470-484

[3] Anser MK, Khan MA, Awan U, Batool R, Zaman K, Imran M, Sasmoko, Indrianti Y, Khan A, Bakar ZA (2020a) The role of technological innovation in a dynamic model of the environmental supply chain curve: Evidence from a panel of 102 countries. PRO 8(9):1033. https://doi.org/10.3390/pr8091033

[4] Anser MK, Yousaf Z, Zaman K. Green technology acceptance model and green logistics operations: "To see which way the wind is blowing". Front Sustain. 2020b;**1**:3 https://doi.org/10.3389/frsus

[5] Kim Y, Rhee DE. Do stringent environmental regulations attract foreign direct investment in developing countries? Evidence on the "race to the top" from cross-country panel data. Emerging Markets Finance and Trade. 2019;**55**(12):2796-2808

[6] Munir IU, Yue S, Nassani AA, Abro MMQ, Hyder S, Zaman K. Structural changes, financial and business regulatory measures, energy and tourism demand: Evidence from a group of seven countries. International Journal of Finance and Economics. 2020. DOI: https://doi.org/10.1002/ijfe.1901

[7] Wang Z, Zhu Y. Do energy technology innovations contribute to CO2 emissions abatement? A spatial perspective. Sci Total Environ. 2020;**726**:138574

[8] Shahnaz R, Shabani ZD. Do renewable energy production spillovers matter in the EU? Renewable Energy. 2020;**150**:786-796

[9] Nassani AA, Awan U, Zaman K, Hyder S, Aldakhil AM, Abro MMQ. Management of Natural Resources and Material Pricing: Global evidence. Resources Policy. 2019b;**64**:101500

[10] Naz S, Sultan R, Zaman K, Aldakhil AM, Nassani AA, Abro MMQ. Moderating and mediating role of renewable energy consumption, FDI inflows, and economic growth on carbon dioxide emissions: Evidence from robust Least Square estimator. Environmental Science and Pollution Research. 2019;**26**(3):2806-2819

[11] Niu T, Yao X, Shao S, Li D, Wang W. Environmental tax shocks and carbon emissions: An estimated DSGE model. Structural Change and Economic Dynamics. 2018;**47**:9-17

Chapter 2

Phase Equilibria for Carbon Capture and Storage

Catinca Secuianu and Sergiu Sima

Abstract

Carbon dioxide (CO2) is an important material in many industries but is also representing more than 80% of greenhouse gases (GHGs). Anthropogenic carbon dioxide accumulates in the atmosphere through burning fossil fuels (coal, oil, and natural gas) in power plants and energy production facilities, and solid waste, trees, and other biological materials. It is also the result of certain chemical reactions in different industry (e.g., cement and steel industries). Carbon capture and storage (CCS), among other options, is an essential technology for the cost-effective mitigation of anthropogenic CO2 emissions and could contribute approximately 20% to CO2 emission reductions by 2050, as recommended by International Energy Agency (IEA). Although CCS has enormous potential in numerous industries and petroleum refineries due their large CO2 emissions, a significant impediment to its utilization on a large scale remains both operating and capital costs. It is possible to reduce the costs of CCS for the cases where industrial processes generate pure or rich CO2 gas streams, but they are still an obstacle to its implementation. Therefore, significant interest was dedicated to the development of improved sorbents with increased CO2 capacity and/or reduced heat of regeneration. However, recent results show that phase equilibria, transport properties (e.g., viscosity, diffusion coefficients, etc.) and other thermophysical properties (e.g., heat capacity, density, etc.) could have a significant effect on the price of the carbon. In this context, we focused our research on the phase behavior of physical solvents for carbon dioxide capture. We studied the phase behavior of carbon dioxide and different classes of organic substances, to illustrate the functional group effect on the solvent ability to dissolve CO2. In this chapter, we explain the role of phase equilibria in carbon capture and storage. We describe an experimental setup to measure phase equilibria at high-pressures and working procedures for both phase equilibria and critical points. As experiments are usually expensive and very time consuming, we present briefly basic modeling of phase behavior using cubic equations of state. Phase diagrams for binary systems at high-pressures and their construction are explained. Several examples of phase behavior of carbon dioxide + different classes of organic substances binary systems at high-pressures with potential role in CCS are shown. Predictions of the global phase diagrams with different models are compared with experimental literature data.

Keywords: phase equilibria, high-pressures, CCS, equations of state, carbon dioxide, phase diagrams

1. Introduction

The worldwide anthropogenic carbon dioxide (CO_2) releases in atmosphere, accounting for over 80% of greenhouse gases (GHGs: > 81% CO_2, 10% methane,

7% nitrous oxide, ~3% fluorinated gases [1, 2]), are emitted through numerous processes such as burning solid waste, trees, other biological materials, and especially fossil fuels (coal, oil, and natural gas) in power plants and energy production facilities, land use change, worldwide bushfires, and also as a result of certain chemical reactions in different industries (e.g., cement and steel factories) [3–5].

In addition, the Mauna Loa Observatory, Hawaii, USA forecast that the annual CO2 concentration will continue to rise in 2020, exceeding for the first time 415 ppm monthly [6], after the historical negative record of 415.39 ppm in May 2019. The higher and higher concentrations of CO2 together with methane, nitrous oxide, and fluorinated gases into atmosphere trap heat which causes the greenhouse effect producing the global warming and the climate change, triggering an escalation in the number and strength of natural disasters such tornados, floods, droughts, melting of ice from both poles, melting of mountain glaciers, and an increase in average sea levels [7, 8].

Among the many ways for carbon mitigation [9, 10], such as wind, solar, or hydro (waves) based electricity generation [4], waste management [11], chemical conversion of carbon [5], etc., one of the most mature technologies is the carbon capture and storage (CCS), which could contribute approximately 20% to CO2 emission reductions by 2050, as recommended by International Energy Agency (IEA) [12].

A key obstacle in the utilization on a large scale of CCS remain the operating and capital costs despite its great potential in many industries and petroleum refineries given their large CO_2 emissions [13, 14]. In the cases of industrial processes that generate pure or rich CO_2 gas streams the costs of CCS could be reduced but they are still an impediment to its deployment [13, 14].

In the last decade, many researchers focused almost exclusively on the development of improved sorbents with increased CO_2 capacity and/or reduced heat of regeneration [14]. However, recent results [3, 5] show that whilst equilibrium CO_2 capacity is a key determinant of process performance, phase equilibria, transport properties (e.g., viscosity, diffusion coefficients, etc.) and other thermophysical properties (e.g., heat capacity, density, etc.) have a significant effect on the capital cost, and thus on the price of the carbon captured.

In this context we recently focused on the phase behavior research of physical solvents for carbon dioxide capture. Phase equilibria (PE) at high-pressures (HP) of carbon dioxide with different classes of organic substances, such as alkanes [15], cycloalkanes [16–18], ethers [19, 20], alcohols [21–24], esters, ketones, were investigated to illustrate the functional group effect on the solvent ability to dissolve CO_2. As the experiments are usually expensive and very time consuming, equations of state (EoS) models are the most common approach for the correlation and/or prediction of phase equilibria and properties of the mixtures [25]. In this chapter we will also show modeling results with the General Equation of State (GEOS) [26, 27], Peng–Robinson (PR) [28] and Soave–Redlich–Kwong (SRK) [29] equations of state (EoS) coupled with both classical van der Waals (two-parameter conventional mixing rule, 2PCMR). Our approach is rather predictive, and we are interested in the global phase behavior of the systems, not only on pressure-composition diagrams, for instance. Therefore, phase diagrams construction is also explained.

2. Experimental method and setup for measuring PE at HP

One of our laboratory experimental setup for measuring phase equilibria at high pressure consists of three modules: equilibrium cell, sampling and analyzing systems and control systems (**Figure 1**). The main component of the experimental apparatus is the high-pressure windowed cell with variable volume [30–32],

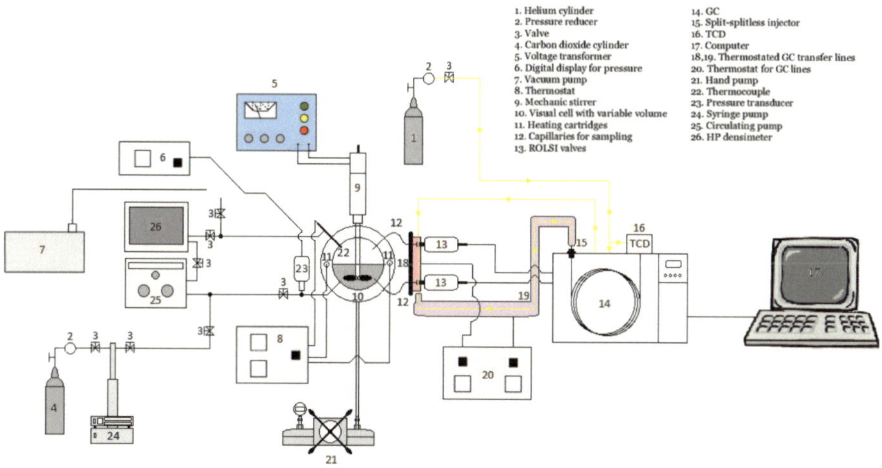

Figure 1.
Schematic diagram of phase equilibria setup [36].

coupled with the sampling and analyzing system [33, 34]. The sampling system consists of two high-pressure electromechanical sampling valves, namely the rapid on-line sampler injector (ROLSI™, MINES ParisTech/CEP-TEP – Centre énergétique et procédés, Fontainbleau, France [35]). The ROLSI valves are connected to the equilibrium cell and to a gas chromatograph (GC) through capillaries. The expansion chamber of the sampler injector is heated with a heating resistance, so the liquid samples are rapidly vaporized. A linear resistor coupled to an Armines/CEP/TEP regulator is used to heat the transferring lines between ROLSI and the GC. The GC (Perichrom) is equipped with a thermal conductivity detector, TCD, and a HP-Plot/Q column 30 m long and 0.530 mm diameter. Helium is the GC carrier gas at a flow rate of 30 mL/min. The setup is completed with a syringe pump Teledyne ISCO model 500D.

The working procedure to determine isothermal phase equilibria is briefly described. Firstly, the entire internal loop of the apparatus including the equilibrium cell is flushed several times with carbon dioxide. Any trace of carbon dioxide is then evacuated from the cell and lines with a vacuum pump. The next step is to load the organic substance in the equilibrium cell, which is previously degassed by using a vacuum pump and vigorously stirring. The lighter component (in this case CO2) is also filled with the syringe pump into equilibrium cell and the pressure is set to the desired value. The experiment continues by heating the cell to the required temperature. The mixture in the cell is stirred for a few hours to facilitate the approach to an equilibrium state. Then the stirrer is switched off for about one hour allowing complete separation of the coexisting phases. Samples from both liquid and vapor phases are withdrawn by ROLSI valves and analyzed with the GC. To check the repeatability, at least six samples of the liquid phase are normally analyzed at the equilibrium temperature and pressure. The sample sizes being very small, the equilibrium pressure in the cell remains constant.

The calibration of the TCD for CO2 and organic substance is done by injecting known amounts of each component using gas chromatographic syringes. Calibration data are fitted to quadratic polynomials to obtain the mole number of the component versus chromatographic area. Then correlation coefficients of the GC calibration curves are calculated.

The uncertainties all variables and properties are estimated [21]. For the phase equilibrium compositions, the relative uncertainty of the mole fraction in the liquid and vapor phases are calculated using the procedure given by Scheidgen [37] and Chirico et al. [38].

A different working procedure is used when measuring critical points of systems at high-pressures. Thus, we start with a homogenous phase with unknown composition at random temperature and pressure. The pressure is modified by varying the volume of the cell with the manual pump to determine if we obtain a bubble or a dew point. If we obtain a bubble point, the temperature is slowly increased until the first dew point is observed, then the pressure is increased to a homogeneous phase and the composition is determined by sampling. Then the pressure is very slowly decreased until the first drops of liquid are observed. At this point, the temperature is slowly decreased simultaneously with reducing the volume, so the system is at the limit between homogeneous (single phase)-heterogeneous (two phases). The decreasing of temperature continues until the first gas bubbles are observed. The procedure is then repeated by introducing new amounts of CO2 and slowly cooling.

3. Modeling

Equations of state (EoS) models are the most common approach for the correlation and prediction of phase equilibria and properties of the mixtures. Cubic and generalized van der Waals equations of state were actively studied, since van der Waals proposed his famous equation in 1873 [39]. Although the cubic EoSs have their known limitations [39–41], they are frequently used for practical applications. They offer the best balance between accuracy, simplicity, reliability, and speed of computation [39–41]. Therefore, cubic equations of state remain important and easy tools to calculate the phase behavior in many systems [40], even for complex mixtures like petroleum fluids [25].

The general cubic equation of state (GEOS) has the form:

$$P = \frac{RT}{V-b} - \frac{a(T)}{(V-d)^2 + c} \tag{1}$$

The four parameters a, b, c, d for a pure component are expressed by:

$$\begin{cases} a(T) = a_c \beta(T_r) a_c = \Omega_a \dfrac{R^2 T_c^2}{P_c} \\ b = \Omega_b \dfrac{RT_c}{P_c} \\ c = \Omega_c \dfrac{RT_c^2}{P_c^2} \\ d = \Omega_d \dfrac{RT_c}{P_c} \end{cases} \tag{2}$$

The temperature function used is:

$$\beta(T_r) = T_r^{-m} \tag{3}$$

with the reduced temperature $T_r = T/T_c$. The expressions of the parameters Ω_a, Ω_b, Ω_c, and Ω_d are:

$$\Omega_a = (1-B)^3; \quad \Omega_b = Z_c - B; \quad \Omega_c = (1-B)^2(B-0.25); \quad \Omega_d = Z_c - \frac{(1-B)}{2} \tag{4}$$

with

$$B = \frac{1+m}{\alpha_c + m}, \tag{5}$$

where α_c is the Riedel's criterion.

It can be easily noticed that a, b, c, d coefficients of the cubic GEOS equation are in fact functions of the critical data (T_c, P_c, and V_c), m, and α_c parameters.

The cubic GEOS equation is a general form for all the cubic equations of state with two, three, and four parameters [42]. This is the meaning of the statement cubic "general equation of state" used for GEOS. To obtain the parameters of the Peng–Robinson (PR) or Soave–Redlich–Kwong (SRK) equations of state from the Eqs. (2–5) we set the following restrictions [42]: $\Omega_c = -2(\Omega_b)^2$and $\Omega_d = -\Omega_b$ or $\Omega_c = -\left(\frac{\Omega_b}{2}\right)^2$ and $\Omega_d = -\frac{\Omega_b}{2}$ respectively. It results:

$$B = 0.25 - \frac{1}{8}\left(\frac{1-3B}{1-B}\right)^2; \quad Z_c(PR) = \frac{1+B}{4.}$$

or

$$B = 0.25 - \frac{1}{36}\left(\frac{1-3B}{1-B}\right)^2; \quad Z_c(SRK) = \frac{1}{3} \tag{6}$$

giving $B(PR) = 0.2296$ and $Z_c(PR) = 0.3074$ or $B(SRK) = 0.2467$.

The equations for B can be solved iteratively, starting with an initial approximation of B in the right hand term. The corresponding value for Ω_a, Ω_b, Ω_c, Ω_d are given by Eq. (4). The Ω_a, Ω_b, Ω_c, Ω_d parameters are compound dependent for GEOS, while for SRK and PR are universal. It must be also pointed out that in GEOS the value of Zc is the experimental one.

The coefficients a, b, c, d were obtained for mixtures using the classical van der Waals two parameters conventional mixing rules (2PCMR for a, b) extended correspondingly for c and d:

$$a = \sum_i \sum_j x_i x_j a_{ij}; \; b = \sum_i \sum_j x_i x_j b_{ij}; \; c = \sum_i \sum_j x_i x_j c_{ij}; \; d = \sum_i x_i d_i \tag{7}$$

$$a_{ij} = \left(a_i a_j\right)^{1/2}\left(1 - k_{ij}\right); \quad b_{ij} = \frac{b_i + b_j}{2}\left(1 - l_{ij}\right) \tag{8}$$

$$c_{ij} = \pm\left(|c_i||c_j|\right)^{1/2} \text{ (with " + " for } c_i, c_j > 0 \text{ and "–" for } c_i \text{ or } c_j < 0\text{)}.$$

Generally, negative values are common for the c parameter of pure components.

The GEOS parameters m and α_c of each component were estimated by constraining the EoS to reproduce the experimental vapor pressure and liquid volume on the saturation curve between the triple point and the critical point.

The phase equilibrium is represented using the Gibbs energy minimization criterion, which can be equivalently written as iso-fugacity criterion [43]:

$$f_i^L = f_i^V \tag{9}$$

where f is the fugacity of the component i in the liquid (L) or vapor (V) phase. Using the fugacity model for both phases, this equation becomes:

$$x_i\phi_i^L(T,p,x_1,\dots,x_N)=y_i\phi_i^V(T,p,y_1,\dots,y_N) \tag{10}$$

where ϕ_i^L and ϕ_i^V are the fugacity coefficients of component i in the liquid and vapor phase respectively. Using the equilibrium constant, $K_i=\frac{y_i}{x_i}$, Eq. (10) becomes:

$$K_i=\frac{\phi_i^L(T,p,x_1,\dots,x_N)}{\phi_i^V(T,p,y_1,\dots,y_N)} \tag{11}$$

The fugacity coefficient of component *i* in the mixture is calculated from any equation of state.

The vapor–liquid equilibrium calculation for *bubble p* or *dew p* type problems involves solving the following system of nonlinear equations:

$$f(T\,or\,p)=\sum_i K_ix_i-1=0 \tag{12}$$

$$y_i=K_ix_i;\sum_i y_i=1 \tag{13}$$

or

$$f(T\,or\,p)=\sum_i\frac{y_i}{K_i}-1=0 \tag{14}$$

$$x_i=\frac{y_i}{K_i};\quad \sum_i x_i=1 \tag{15}$$

respectively.

The calculations were made using the software package *PHEQ* (Phase Equilibria Database and Applications), developed in our laboratory [44], and the GPEC (Global Phase Equilibrium Calculations) [45]. In our in-house software (CRIMIX), the critical curves are calculated using the method proposed by Heidemann and Khalil [46] with the numerical derivatives given by Stockfleth and Dohrn [47].

4. Phase diagrams

The phase diagram shows the domains occupied by the different phases of a system, the boundaries that separate these regions and the special points of the system, as a function of two independent variables. A practical choice of these variables is that of pressure (*P*) and temperature (*T*), which can be measured experimentally. The phase law states that for a one-component system, the coexistence curves, whether liquid-vapor, solid-vapor or solid–liquid, are monovariants. The curves separate in the phase diagram the domains of existence of each phase (vapor, liquid, solid).

The most known classification of phase diagram types was proposed by Scott and van Konyneburg [48]. They applied the van der Waals equation combined with van der Waals-type mixing rules in binary systems and quantitatively predicted almost all types of phase equilibria of fluids known from experiments. The results were presented in a global phase diagram and its pressure–temperature (*P–T*) projections. Scott and van Konyneburg [48] classified the different types of phase diagrams taking into account the nature of *P–T* projections and, in particular, the presence or absence of three-phase lines and azeotropic lines, as well as how critical lines connect with these. According to Scott and van Konyneburg's classification, six main types of phase diagrams can be distinguished. The first five types of fluid behavior were calculated by [48] with the above equation, the sixth type being calculated with other equations of state [32]. More recently, Privat and Jaubert [49], presented an updated version of classification of the phase diagrams in binary systems together with the transitions between the various types of systems. The six main types of phase diagrams are as follows (**Figure 2** [32]):

- Type I – presents a critical liquid-vapor curve which connects the critical points of the two pure components;

- Type II – presents two critical curves. It has a critical curve as in the case of type I and, in addition, a liquid–liquid critical curve that starts at an upper critical endpoint (UCEP) and evolves rapidly at high pressures and a three-phase liquid–liquid-vapor (LLV) equilibrium curve ending in UCEP;

- Type III – presents three critical curves. A critical liquid-vapor curve that starts at the critical point of one of the pure components and ends in a UCEP where the curve of the three liquid–liquid-vapor phases ends; another liquid-vapor critical curve starting from the critical point of the other pure component and connecting with a liquid–liquid critical curve at very high pressures;

- Type IV – presents two distinct liquid-vapor critical curves starting from each of the critical points of the pure components and a liquid–liquid–vapor equilibrium curve of the three phases bordering the two liquid–vapor critical curves, the two critical points where it intersects representing a UCEP and, respectively, a lower critical endpoint (LCEP). Also, in this type of diagram, a liquid–liquid critical curve is found at low temperatures, ending in another UCEP, in which an equilibrium curve of the three LLV phases ends;

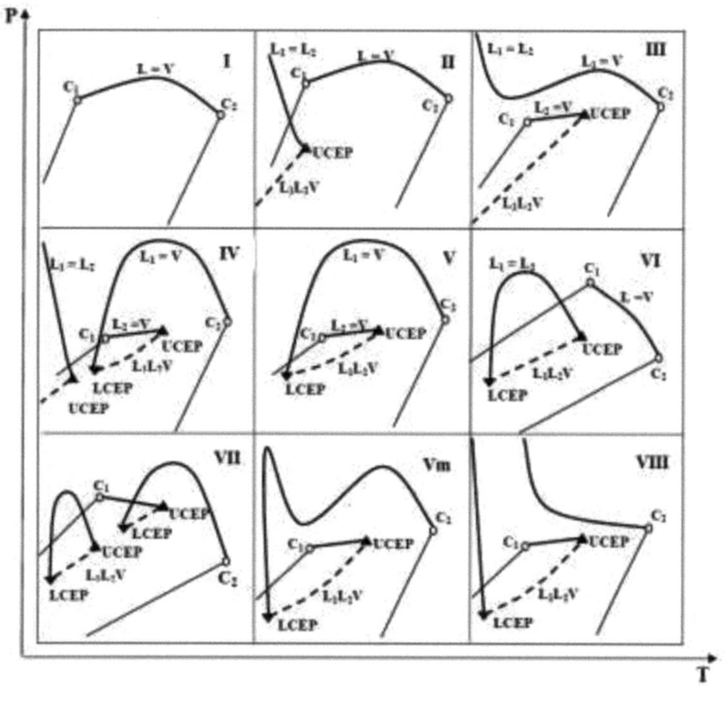

Figure 2.
Schematic representation of the six types of phase diagrams [32].

- Type V – presents a critical curve similar to that of type IV, but without the liquid–liquid critical curve and without the equilibrium curve of the three-phases at low temperatures;

- Type VI – presents a critical curve as in the case of type I, but also a critical liquid–liquid curve that stretches between an LCEP and a UCEP, which also border the equilibrium curve of the three phases.

Subsequently, the classification of phase diagrams was enriched with several new types of critical curves [32]:

- Type VII - this phase diagram shows the same type of liquid–liquid critical curve as type VI, but the liquid-vapor critical curve is composed of two sections ending in a UCEP, respectively in an LCEP, which in turn border a liquid–liquid-vapor curve of the three phases;

- Type VIII -has a critical curve as in the case of type III which additionally presents a liquid–liquid critical line that starts at a lower critical limit point, LCEP and then goes to high pressures [30];

- Type Vm - the critical curve is interrupted by a curve of the three phases and has a maximum and a minimum in pressure.

In the case of binary systems, especially, the sections *P-X* at *T* = constant or *T-X* at *P* = constant are widely used. Such sections are constructed as in **Figures 3** and **4**. In *P–T* projections, the axis of the composition is hidden by the pressure–temperature plane. The vapor pressure curve for component 1 is denoted 1 and for component 2, 2. In **Figure 3**, a *P–T* projection of the type I diagram is presented, but, in addition, for a composition, *X*, we represented the bubble pressure curves (BPC) and dew pressure curves (DPC) that intersects at a point on the critical liquid-vapor curve, where the liquid and vapor phases become identical.

Each critical point on the critical liquid-vapor curve corresponds to a certain composition. The points near the critical point of component 1 are points in the mixture where the concentration of component 1 is very high, while the points located near the component 2 are critical points of the mixture in which its concentration is very high.

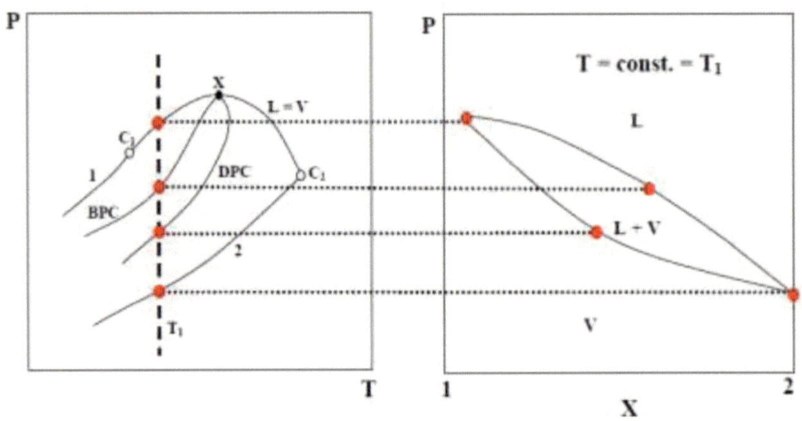

Figure 3.
P–T and P-X projections for type I phase diagram [32].

DOI: http://dx.doi.org/10.5772/intechopen.95136

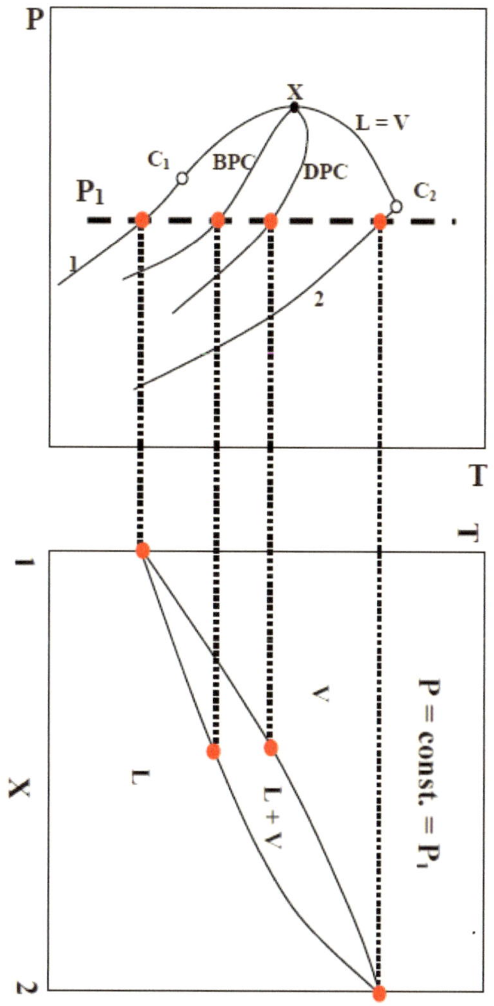

Figure 4.
P–T *and* T-X *projections for type I phase behavior [32].*

To construct a *P-X* diagram at T = constant (**Figure 3**), for example T_1, a line parallel to the pressure axis is drawn. This line, corresponding to the temperature T_1, firstly meets the vapor pressure curve of component 2, then the dew pressure curves (DPC), respectively the boiling pressure curves (BPC) and, finally, the vapor pressure curve of component 1.

All mixtures of components 1 and 2 are represented at low pressure by a single vapor phase (single-phase domain) and at high pressure by a single liquid phase (single-phase domain). The biphasic domain, delimited by the liquid curve, respectively by that of the vapors, corresponds to the equilibrium state between phases (liquid + vapors).

To construct a *T-X* diagram at P = constant (**Figure 4**), one can proceed similarly, but draw a line parallel to the temperature axis.

The complexity of *P–T* and *P-X* projections increases as the phase behavior is more sophisticated. As examples, the *P–T* and *P-X* projections for a type III phase diagram are shown in **Figures 5** and **6** for two constant temperatures: one lower than the critical point of the pure component 1 and the other located between the critical point 1 and the UCEP.

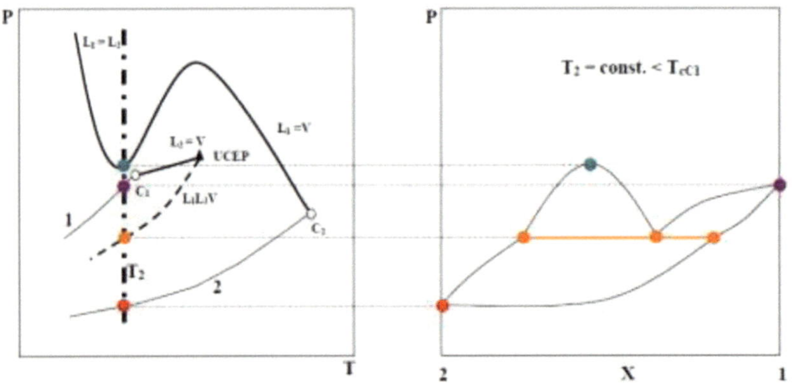

Figure 5.
P–T *and* P-X *projections for type III phase diagram for temperature lower than the temperature of pure component 1 [32].*

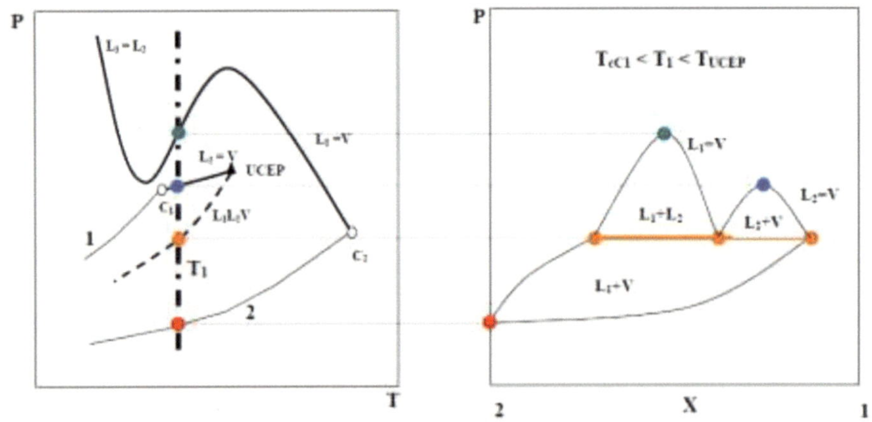

Figure 6.
P–T *and* P-X *projections for type III phase diagram for temperature higher than the temperature of pure component 1 and lower than the UCEP's temperature [32].*

5. Predicting the phase behavior of systems containing carbon dioxide of interest for CCS

Binary systems containing carbon dioxide and organic substances are of interest for CCS both from fundamental point of view and for the ability of different substances to dissolve CO_2. Different classes of organic substances are studied to explore the functional group effect on the solvent ability to dissolve CO_2. Several examples are discussed in the next sections.

5.1 Phase diagram predictions for the carbon dioxide +1-dodecanol binary system

As explained in previous section, the phase diagrams can be calculated using equations of state. The experimental data can be correlated at each temperature and sets (k_{ij}, l_{ij}) of optimized binary interaction parameters (BIPs) are obtained. Each set then can be used to calculate the phase diagram of that system generating as many diagrams as the temperatures used.

One example is shown in **Figure 7** where for the carbon dioxide +1-dodecanol binary system [22] are plotted the critical curves calculated with the PR/2PCMR model using the optimized binary interaction parameters obtained at 293.15 K (k_{12} = 0.1277; l_{12} = −0.0026), 303.15 K (k_{12} = 0.1266; l_{12} = −0.0068), 313.15 K (k_{12} = 0.1234; l_{12} = −0.0055), 333.15 K (k_{12} = 0.1535; l_{12} = −0.0193), and 353.15 K (k_{12} = 0.1380; l_{12} = −0.0100).

Another possibility is to use one constant set of BIP to calculate the phase diagram. In the same figure, the critical curve (dark red continuous line) of the system is also calculated with a constant set of BIPS obtained by the averaging the optimized BIPs for the five temperatures mentioned (k_{12} = 0.1278; l_{12} = −0.0006).

The calculation with the average value of BIPs is not representing well the experimental critical data (blue full circles), therefore a modified set of BIPs (k_{12} = 0.0900; l_{12} = −0.0100) was obtained by a trial and error procedure [22]. This set of parameters is predicting reasonably well the type phase behavior of the carbon dioxide +1-dodecanol binary system.

5.2 Phase diagram predictions for carbon dioxide + different classes of substances at high-pressures

A more severe test for an equation of state is to use the BIPs obtained for a mixture to calculate the phase diagram for other ones. **Figure 8** illustrates the predictions of phase diagrams for five binary systems containing carbon dioxide and different classes of organic substances containing four C atoms: one normal alkane (n-butane, n-C_4H_{10}), one primary (1-butanol, $C_4H_{10}O$) and one secondary alcohol (2-butanol, $C_4H_{10}O$), one ester (ethyl acetate, $C_4H_8O_2$), and one ether (1,2-dimethoxyethane (1,2-DME), $C_4H_{10}O_2$).

The calculations were made with the SRK/2PCMR EoS. In all cases, the model predicts type II phase behavior, even if only for carbon dioxide +1-butanol and carbon dioxide +2-butanol there is experimental evidence that they exhibit liquid–liquid immiscibility [50]. One unique set of binary interaction parameters tailored for the carbon dioxide +2-butanol system was used to predict the phase behavior of

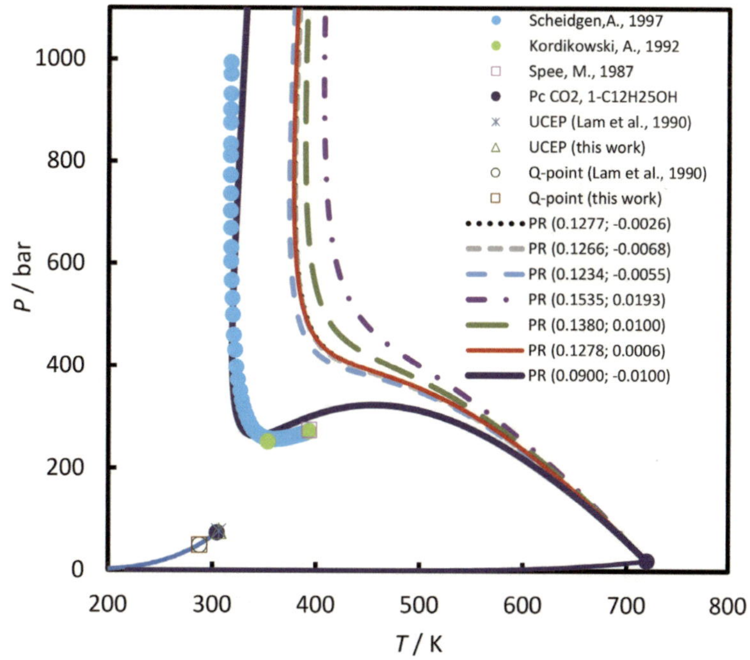

Figure 7.
P–T *fluid phase diagram of carbon dioxide (1) + 1-dodecanol (2) system [22].*

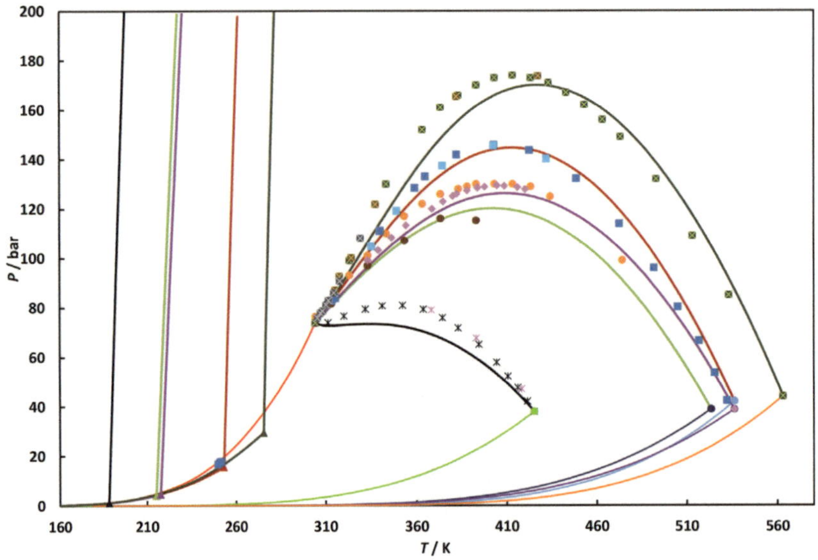

Figure 8.
P–T *fluid phase diagrams of carbon dioxide (1) + n-butane (2), + ethyl acetate (2), + 1,2-dimethoxyethane (2), + 2-butanol (2), + 1-butanol (2) binary systems. Symbols, experimental data; lines, predictions by SRK/2PCMR* (k_{12} = 0.0200; l_{12} = −0.1110).

the other four binary systems. Thus, SRK EoS was firstly used in a semi-predictive approach to obtain a set of binary parameters yielding good results for the binary system carbon dioxide +2-butanol (including VLE in the entire temperature range, critical curve). The set of binary parameters was calculated using the k_{12}-l_{12} method [50] to obtain the experimental value of the vapor-liquid critical pressure maximum (CPM) simultaneously with the temperature of UCEP. The experimental temperature of UCEP and CPM have been traced by paths in the k_{12}-l_{12} diagram and their intersection has given the values of the parameters [50]. The values of the interaction binary parameters fulfilling these requirements are k_{12} = 0.0200 and l_{12} = −0.1110. This set of BIPs was used then to predict the phase behavior for the carbon dioxide + n-butane (black curve), + ethyl acetate (green curve), + 1,2-DME (pink curve), and + 1-butanol (olive curve). Except for the carbon dioxide + n-butane binary system, the critical curves are remarkably well predicted for all mixtures. It should be noted that for the carbon dioxide + ethyl acetate system the experimental critical data are scattered (orange and brown full circles), while for the carbon dioxide +1-butanol system the model predicts the CPM at a higher temperature than the experimental one. The predicted UCEPs are increasing in temperature in the order: $T_{CO2\ +\ n\text{-}butane} < T_{CO2\ +\ ethyl\ acetate} < T_{CO2\ +\ 1,2\text{-}DME} < T_{CO2\ +\ 2\text{-}butanol} < T_{CO2\ +\ 1\text{-}butanol}$.

The critical lines from **Figure 8** indicate that the binary of carbon dioxide +1-butanol displays the larger range of immiscibility, followed by carbon dioxide +2-butanol, carbon dioxide +1,2-DME, and carbon dioxide + ethyl acetate, compared with the binary of carbon dioxide + n-butane, the corresponding n-alkane with the same carbon number.

5.3 Comparisons of phase behavior with different models

The carbon dioxide +2-butanol binary system can be used as a model system as experimental data are available for the entire range of critical temperatures and pressures, including the UCEP [50]. **Figure 9** shows the predicted critical lines for the carbon dioxide +2-butanol system with three models: GEOS, SRK, and PR using a single set of BIPs for each model [50]. The BIPs were obtained using the k_{12}-l_{12} method explained in previous sub-section. It can be observed that all models predict very well the liquid-vapor critical curve. In addition, GEOS equation reproduces very well the CPM and the corresponding temperature, while SRK and PR predict a similar value for the CPM but shifted to a higher temperature. The models show a comparable behavior for the liquid–liquid critical curves, the only difference being the predicted UCEPs. However, the models show bigger difference in the critical pressure-compositions and critical temperature compositions projections of the phase diagrams, as can be seen in **Figure 10**. The prediction results by GEOS are

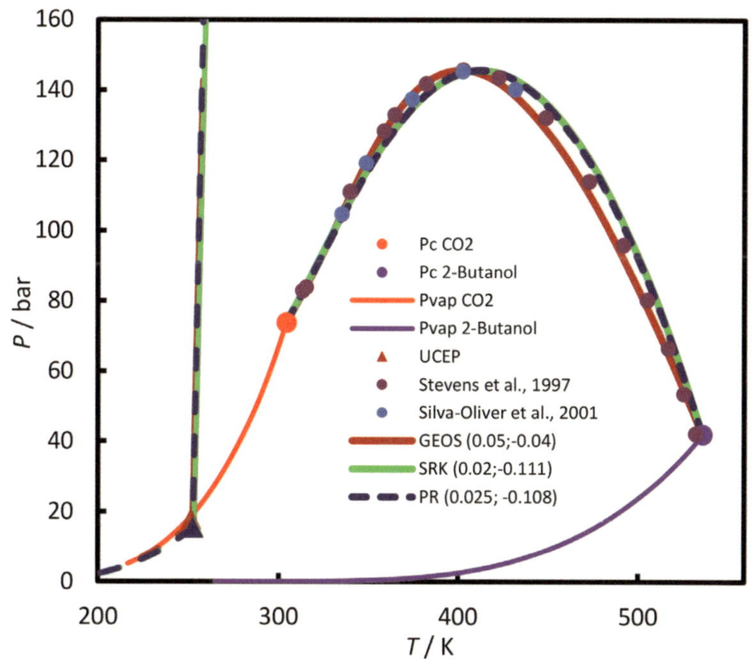

Figure 9.
P–T *fluid phase diagrams of carbon dioxide (1) + 2-butanol (2) predicted by GEOS, SRK, and PR models.*

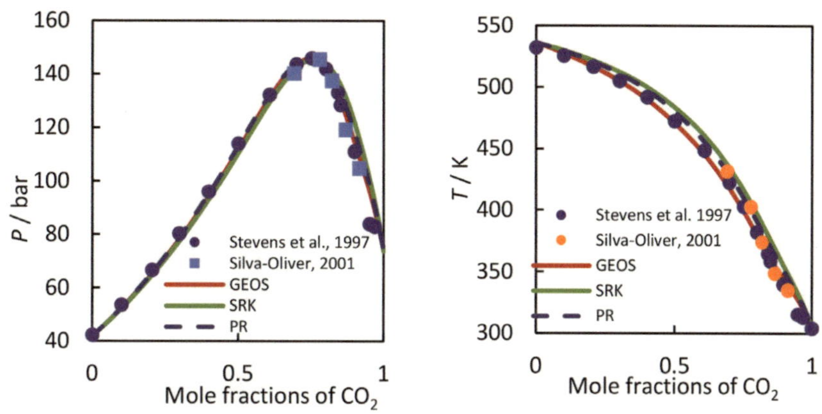

Figure 10.
P–X$_1$ *and* T-X$_1$ *projections of the critical curve of carbon dioxide (1) + 2-butanol (2) predicted by GEOS, SRK, and PR models.*

better than PR and SRK respectively. The differences in the predictions will than be reflected in the prediction of the bubble-point and dew-point curves [50].

6. Conclusions

The phase behavior of selected hydrocarbons and functional groups substances (alcohols, ethers, esters) is investigated to assess their capability as solvents/cosolvents for CO2 capture. Simple models as cubic equations of state have the capability to accurately predict the phase behavior of carbon dioxide + different organic substances binary systems. Phase diagrams are important tools in assessing the global behavior. The type of phase behavior is providing essential information for the operating domain of binary system.

Acknowledgements

This work was supported by a grant of Ministry of Research and Innovation, CNCS - UEFISCDI, project number PN-III-P4-ID-PCE-2016-0629, within PNCDI III.

Conflict of interest

The authors declare no conflict of interest.

Author details

Catinca Secuianu[1,2]* and Sergiu Sima[1]

1 Department of Inorganic Chemistry, Physical Chemistry and Electrochemistry, Faculty of Applied Chemistry and Materials Science, University Politehnica of Bucharest, Bucharest, Romania

2 Department of Chemical Engineering, Imperial College London, South Kensington Campus, London, United Kingdom

*Address all correspondence to: catinca.secuianu@upb.ro

References

[1] IEA, Global Energy Review 2020, IEA, Paris. 2020. Available from: https://www.iea.org/reports/global-energy-review-2020. [Accessed: 2020-05-10]

[2] Greenhouse Gas Emissions, 2020, Available from: https://www.epa.gov/ghgemissions/overview-greenhouse-gases [Accessed: 2020-08-05]

[3] Zheng J, Chong ZR, Fahed Qureshi M, Linga P: Carbon Dioxide Sequestration via Gas Hydrates: A Potential Pathway toward Decarbonization. Energy Fuels. 2020;34: 10529–10546. DOI: https://dx.doi.org/10.1021/acs.energyfuels.0c02309

[4] Cho HH, Strezov V: A Comparative Review on the Environmental Impacts of Combustion-Based Electricity Generation Technologies. Energy Fuels. 2020;34: 10486–10502. DOI: https://dx.doi.org/10.1021/acs.energyfuels.0c02139

[5] D_Alessandro DM, Smit B, Long JR: Carbon Dioxide Capture: Prospects for New Materials. Angew. Chem. Int. Ed. 2010;49:6058–6082. DOI: 10.1002/anie.201000431

[6] Trends in Atmospheric Carbon Dioxide. 2020. Available from: https://www.esrl.noaa.gov/gmd/ccgg/trends/ [Accessed 2020-05-05]

[7] Haszeldine RS: Carbon Capture and Storage: How Green Can Black Be? Science. 2009;325:1647–1652. DOI: 10.1126/science.1172246

[8] Kenarsari SD, Yang D, Jiang G, Zhang S, Wang J, Russell AG, Wei Q, Fan M: Review of recent advances in carbon dioxide separation and capture. RSC Adv. 2013;3:22739–22773. DOI: 10.1039/c3ra43965h

[9] S. A. R. Khan, Y. Zhang, A. Kumar, E. Zavadskas, D. Streimikiene: Measuring the impact of renewable energy, public health expenditure, logistics, and environmental performance on sustainable economic growth. Sustainable Development. 2020;28:833–843. DOI: 10.1002/sd.2034

[10] S. A. R. Khan, A. Sharif, H. Golpîra, A. Kumar: A green ideology in Asian emerging economies: From environmental policy and sustainable development. Sustainable Development. 2019;27:1063–1075. DOI: 10.1002/sd.1958

[11] Buck HJ: Should carbon removal be treated as waste management? Lessons from the cultural history of waste. Interface Focus. 2020;10: 20200010. DOI: http://dx.doi.org/10.1098/rsfs.2020.0010

[12] IPCC, Climate Change 2021: Mitigation of Climate Change: Contribution of Working Group III to the Sixth Assessment Report of the Intergovernmental Panel on Climate Change. 2019. Available from: https://www.ipcc.ch/report/sixth-assessment-report-working-group-3/ [Accessed: 2020-06-15]

[13] Mota-Martinez MT, Hallett JP, Niall Mac Dowell N: Solvent selection and design for CO2 capture – how we might have been missing the point. Sustainable Energy Fuels. 2017; 1:2078–2090. DOI: 10.1039/c7se00404d

[14] Bui M, Fajardy M, Mac Dowell N: Bio-energy with carbon capture and storage (BECCS): Opportunities for performance improvement. Fuel. 2018; 213: 164–175. http://dx.doi.org/10.1016/j.fuel.2017.10.100

[15] Secuianu C, Feroiu V, Geană D: Phase Behavior for the Carbon Dioxide plus N-Pentadecane Binary System. Journal of Chemical and Engineering

Data. 2010;55: 4255–4259. DOI: 10.1021/je100404g

[16] Sima S, Cruz-Doblas J, Cismondi M, Secuianu C: High-pressure phase equilibrium calculations for carbon dioxide plus cyclopentane binary system. Central European Journal of Chemistry. 2014;9: 918–927. DOI: 10.2478/s11532-013-0393-2

[17] Sima S, Milanesio JM, Ramello JI, Cismondi M, Secuianu C, Feroiu V, Geană D: The effect of the naphthenic ring on the VLE of (carbon dioxide plus alkane) mixtures. Journal of Chemical Thermodynamics. 2016;93: 374–385. DOI: 10.1016/j.jct.2015.07.018

[18] Crisciu A, Sima S, Deaconu AS, Chirila A, Deaconu D, Secuianu C, Feroiu V: Modelling of the Carbon Dioxide plus Cyclohexane Binary System with Cubic Equations of State. Revista de Chimie. 2016;67:1984–1989.

[19] Sima S, Secuianu C, Feroiu V: Phase equilibria of CO2+1,2-dimethoxyethane at high-pressures. Fluid Phase Equilibria. 2018;458:47–57. DOI: 10.1016/j.fluid.2017.11.008.

[20] Sima S, Racovita RC, Chirila A, Deaconu D, Feroiu V, Secuianu C: Phase behaviour calculations for the carbon dioxide+1,2-dimethoxyethane binary system with a cubic equation of state. Studia Universitatis Babes-Bolyai Chemia. 2019;64:129–142. DOI: 10.24193/subbchem.2019.3.11

[21] Secuianu C, Ionita S, Feroiu V, Geană D: High pressures phase equilibria of (carbon dioxide+1-undecanol) system and their potential role in carbon capture and storage. Journal of Chemical Thermodynamics. 2016;93:360–373. DOI: 10.1016/j.jct.2015.08.005

[22] Secuianu C, Feroiu V, Geană D: Phase behavior of the carbon dioxide+1-dodecanol system at high pressures. Fluid Phase Equilibria. 2016;428:62–75. DOI: 10.1016/j.fluid.2016.06.014

[23] Sima S, Ionita S, Secuianu C, Feroiu V, Geană D: High-Pressure Phase Equilibria of Carbon Dioxide+1-Octanol Binary System. Journal of Chemical and Engineering Data. 2018;63:1109–1122. DOI: 10.1021/acs.jced.7b00865

[24] Sima S, Secuianu C, Feroiu V, Ionita S, Geană D: High-pressure phase equilibria of carbon dioxide+2-octanol binary system. Fluid Phase Equilibria. 2020; 510: 112487. DOI: 10.1016/j.fluid.2020.112487

[25] Secuianu C, Qian JW, Privat R, Jaubert JN: Fluid Phase Equilibria Correlation for Carbon Dioxide+1-Heptanol System with Cubic Equations of State. Industrial & Engineering Chemistry Research. 2012;51:11284–11293. DOI: 10.1021/ie3015186

[26] Geană D: A new equation of state for fluids. I. Applications to PVT calculations for pure fluids. Rev. Chim. (Bucharest). 1986;37:303–309.

[27] Geană D: A new equation of state for fluids. II. Applications to phase equilibria. Rev. Chim. (Bucharest). 1986;37:951–959.

[28] Peng D-Y, Robinson DB: A new two-constant equation of state. Ind ustrial and Engineering Chemistry Fundamentals. 1976;15:59–64, DOI: 10.1021/i160057a011

[29] Soave G: Equilibrium constants from a modified Redlich-Kwong equation of state. Chemical Engineering Science. 1972;27:1197–1203. DOI: 0.1016/0009–2509(72)80096–4

[30] Secuianu C, Feroiu V, Geană D: High-pressure vapor–liquid equilibria in the system carbon dioxide and 2-propanol at temperatures from 293.25 K to 323.15 K. Journal of Chemical and

Engineering Data. 2003;48:1384–1386. DOI: 10.1021/je034027k

[31] Secuianu C, Feroiu V, Geană D:High-pressure phase equilibria for the carbon dioxide + methanol and carbon dioxide + isopropanol systems. Rev. Chim.-Bucharest. 2003;54:874–879

[32] Secuianu C: High Pressure Vapor–Liquid Equilibria in the Binary Systems Carbon Dioxide + Monohydroxilic Alcohols [PhD thesis] Bucharest: University Politehnica of Bucharest; 2004.

[33] Sima S, Feroiu V, Geană D: New high pressure vapor–liquid equilibrium data and density predictions for the carbon dioxide + ethanol system. Journal of Chemical and Engineering Data. 2011;56:5052–5059. DOI: 0.1021/je2008186

[34] Sima S, Feroiu V, Geană D: New high pressure vapor–liquid equilibrium data and density predictions for carbon dioxide + ethyl acetate system. Fluid Phase Equilibria. 2012;325:45–52. DOI: 10.1016/j.fluid.2012.03.028

[35] Guilbot P, Valtz A, Legendre H, Richon D: Rapid on-line sampler-injector: a reliable tool for HT-HP sampling and on-line GC analysis. Analusis. 2000;28:426–431. DOI: 0.1051/analusis:2000128

[36] Sima S: High-pressures phase equilibria in systems with carbon dioxide [PhD thesis] Bucharest: University Politehnica of Bucharest; 2012.

[37] Scheidgen A: Fluid Phase Equilibria of Binary and Ternary Carbon Dioxide Mixtures with Hardly Volatile Organic Substances up to 100 MPa. Cosolvency Effect, Miscibility Windows and Holes in the Critical Plane [PhD thesis] Bochum: Ruhr-University Bochum; 1997.

[38] Chirico RD, Frenkel M, Diky VV, Marsh KN, Wilhoit RC: ThermoML - An XML-based approach for storage and exchange of experimental and critically evaluated thermophysical and thermochemical property data. 2. Uncertainties. Journal of Chemical and Engineering Data. 2003;48:1344–1359. DOI: 10.1021/je034088i

[39] Anderko A. Cubic and generalized van der waals equations. In: Sengers JV, Kayser RF, Peters CJ, White Jr HJ, editors. Equations of State for Fluids and Fluid Mixtures. Elsevier B.V.; 2000. p. 75–126. DOI: https://doi.org/10.1016/S1874-5644(00)80015-6

[40] Valderrama JO: The State of the Cubic Equations of State. Industrial & Engineering Chemistry Research. 2003; 42:1603–1618. DOI: https://doi.org/10.1021/ie020447b

[41] Kontogeorgis GM, Privat R, Jaubert JN: Taking Another Look at the van der Waals Equation of State–Almost 150 Years Later. Journal of Chemical and Engineering Data. 2019; 64:4619–4637. DOI: https://doi.org/10.1021/acs.jced.9b00264

[42] Geană D, FeroiuV: Thermodynamic properties of pure fluids using the GEOS3C equation of state. Fluid Phase Equilibria. 2000;174:51–68. DOI: 10.1016/S0378-3812(00)00417-9

[43] Geană D, FeroiuV. Equations of State. Phase Equilibria Applications. 1st ed. Bucharest: Editura Tehnică; 2000.

[44] Geană D, Rus L. Phase Equilibria Database and Calculation Program for Pure Components Systems and Mixtures. In: Proceedings of the 14th Romanian International Conference on Chemistry and Chemical Engineering (RICCCE XIV); 7–10 September 2005; Bucharest; 2005;2, p. 170–178

[45] http://gpec.phasety.com

[46] Heidemann RA, Khalil AM: The calculation of critical points. AIChE J. 1980;26:769–779. DOI: 10.1002/aic.690260510

[47] Stockfleth R, Dohrn R: An algorithm for calculating critical points in multicomponent mixtures which can easily be implemented in existing programs to calculate phase equilibria. Fluid Phase Equilibria. 1998;145:43–52. DOI: 16/S0378–3812(97)00225–2

[48] van Konynenburg PH, Scott RL: Philos. Trans. R. Soc. London. Series A. 1980;298:495–540

[49] Privat R, Jaubert JN: Classification of global fluid-phase equilibrium behaviors in binary systems. Chemical Engineering Research and Design. 2013; 91: 1807–1839,DOI: 10.1016/j.cherd.2013.06.026

[50] Secuianu C, Feroiu V, Geana D: Phase Behavior for the Carbon Dioxide +2-Butanol System: Experimental Measurements and Modeling with Cubic Equations of State. Journal of Chemical and Engineering Data. 2009;54:1493–1499. DOI: 10.1021/je800799n

Section 2

Carbon and Environment

Chapter 3

Microwave Caustic Slurry Carbonation of Flue Gas of Coal Power Plants in Double Hot Tube Bed for CO2 Sequestration

Yıldırım İsmail Tosun

Abstract

There have been very few transport studies of caustic alkali slurry (metal fines-caustic alkali salt mixture). Bath serpentinite particle size changed the heat conductivity to salt bath. A major reason is that the retention time in fixed film processes is longer than in solid–gas processes. This allows more time to the heat absorption for cracking to the desorbed persistent compounds. Furthermore, heavy serpantinite allows an sufficient intimate contact between coal and biomass surface pores and gas atmosphere in the furnace due to more pyrolysis gas desorption. For seeing the sustainability sequestration and environmental concerns in feasibility sight, the microwave heating technologies encompassing natural carbonation, precipitates for soil remediation and toxic gas sorption was offered to be adopted in Şırnak Asphaltite/Batman Oil Fields cases. In many places, amine sequestration techniques can work synergistically for better results. This study determines to a great extent both the high rate and degree of carbonation under pressurized sludge at 5–10 bar so it was found that, a porous sludge bath over 45% sludge was more efficiently conducted even at a low amount serpantinite slime weight rate, below weight rate of 15%.

Keywords: microwave radiation, waste slurries, metal carbonation, carbonation kinetics, sorbent simulation, hybrid sorbent, waste sludge, salt slurries, char composts, shale compost

1. Introduction

This study was searching possibiliy of CO2 stored in the identification of suitable geological media to and from thermal power plants with installed capacity of over 400 MW in Turkey, cement plants [1] from, the iron and steel industry, sugar involves the calculation of the released CO2 emissions from the refineries and plants [1]. When the emission sources and possible storage locations were evaluated, Silopi Asphaltitel field was selected and it was decided to use the emissions of a cement plant, which is approximately 130 km away from this field. Injection was modeled into the oil field, where both storage and additional oil production could be obtained, accordingly, technical design was proposed for transportation and injection from the CO2 source to the storage area and economic evaluation was made.

There are no CO2 capture facilities in the cement factory, but it was assumed that CO2 was present in the factory area during modeling.

2. Storage of compressed CO2

Carbon dioxide storage underground in Turkey has required to determine the suitable geo storage capacity of the rock structures to be chosen [2]. The goal was to examine the details of the storage of carbon dioxide by injecting it underground in a selected area with geological and numerical models and to make an economic analysis.

The work in this study was carried out in steps:

1. To provide the annual fuel amount and fuel types of thermal power plants, refineries, cement factories, iron and steel industry facilities where carbon dioxide emissions are the highest,

2. Calculation of CO2 emission amounts from fuel consumption according to IGCC method,

3. Oil tank will be in Turkey, natural gas, deep aquifers and examination of structures, such as soda, caves,

4. Modeling the injection and storage of CO2 in the structure to be selected,

5. Technical and economical feasibility analysis of the storage of CO2, which is accepted to be separated by holding technology in the industrial facility, in the field to be selected by using the options of transportation by tanker or pipeline,

6. Technical trips to selected facilities and meetings to introduce the project.

2.1 Investment modeling of compressed CO2 storage from coal and biomass combustion

When the CO2 emission inventory is examined, it is seen that thermal power plants, cement factories, iron and steel industry and refineries are the main sources of CO2 emission in high amounts and in centralized areas. Emissions from transport and domestic use are scattered sources and it is thought that they can be reduced by efficiency measures.

The emission statisticsy in Turkey showed that the high increase in annual emissions [3]. Although it is aimed to increase energy efficiency and the use of renewable energy sources, it is predicted that dependence on fossil fuels will continue in all countries. In this case, another measure that should be taken would be to support technologies that will reduce CO2 emissions from coal-fired thermal power plants. Since CO2 is released as a product in all combustion processes, it is still not possible to eliminate CO2 completely. Therefore, measures should be taken to reduce CO2 emissions. The reduction of CO2 emissions will be possible by underground storage of CO2 in geological structures. Considering that CO2 emissions from industrial zones around the world are determined as 13.5 Gt/year [4], geological areas other than known oil and gas reservoirs should also be evaluated.

The cost of capturing and deposition as below Eq. 1:

$$u(x, \sigma, c) = (1/\sigma)\,(1 + k/\sigma\,(x - \Theta))^{-1-1/k} \tag{1}$$

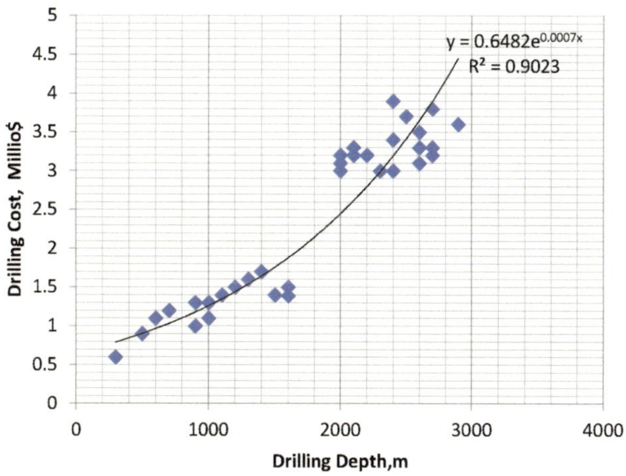

Figure 1.
Change of the cost of drilling for compressed air/CO2 storage reservoir.

while exponential distribution k = 0, where u is the cost function, σ is the variance, k is hybrid distribution parameter, Θ is the time parameter, x is the flow rate by the following **Figure 1** [5].

3. Carbonation of flue gas at microwave column

The oxygen content and electroopotantial of waters adequately accounted in stream flows causing animal feed contamination in the pastoral fields by soil and growing grass nearby this contaminated stream see page [5, 6]. The **Figures 2** and **3** showed the carbonate stability region in the sequestration phase layer and electropotential quality of the sludge.

In the hot streams and acidic mine waters the ferric iron and sulfate tend to be highly common as AMD seepage, alkali resulting from the reduction of these two

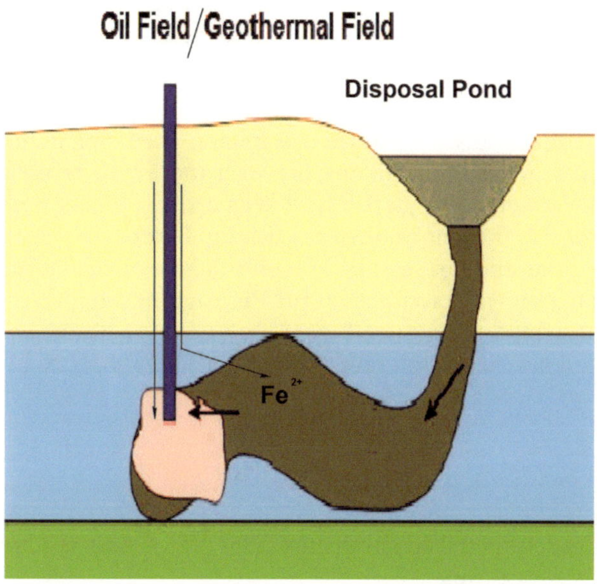

Figure 2.
Storage phase layer diagrams for oil and geothermal field - carbonate stability.

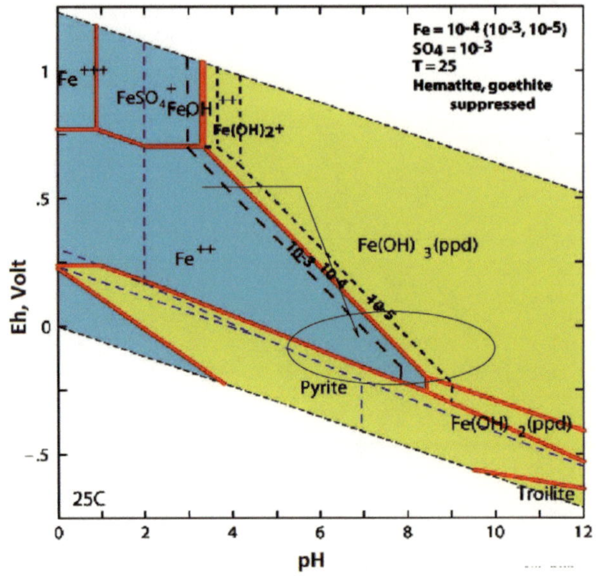

Figure 3.
Eh-pH diagrams for metal carbonate stability.

species, a weak base (bicarbonate) and producing astrong base (hydroxyl ions), also generate net alkalinity. The indirect acid production was relatively high rate at higher pH levels over 5 with dissolution of heavy metals in sulphide minerals and neutralization by alkali matters govern the dissolution by the reactions as given below;

Sludge microwave heating was managed by susceptor ferrous slime addition. The Ca ferrite and serpentinite slime were used and heating temperatures are shown in **Figure 4**.

The serpentinite was waste of chromite mines containing high Fe %23–26 and Mg and Al in silicate form. The waste slime of concentration tables of chromite ores of Elazığ chromites contained 36% iron oxide and provided sufficient heating diagram of filter columns for pressurized sludge column as seen in **Figure 4**.

The aerated and oxygen rich waters oxidizing sulphidic character metal precipitates to sulphate and chloride dissolution by unstable forms, but over ph 9 as shown in Figure electropotantial matter of waste waters provides hydroxide precipitates in soil mud. Even jarosite form precipitates occurring in hot water streams area with reddish brownish precipitates, however those type resıduals stuck over sand may become sweet salty alga fish feed even causing higher heavy metal contamination for fish farming and stream fishing. Batman province copper and lead sulphide deposits and hot streams of high sulphate come out high potential contamination of fresh waters sources at pH Eh diagram stability as given in **Figure 3** following flood.

The dissolution kinetics of soil mud particle for Pb heavy metal is followed by Eq. 2 and 4

$$\frac{dc_{CO2}}{dt} = k_i e^{-tic} dc \quad (2)$$

Where cPb Lead/Fe metal contamination mg/l, k the rate of dissolution of lead, i is the reaction style, t is time,

$$Fe/PbS/PbO + HCO3 = Pb\ SO4 + H2O \quad (3)$$

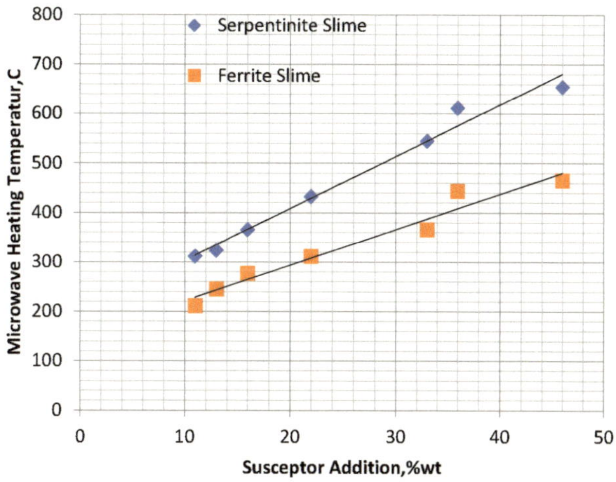

Figure 4.
Carbonation sludge column temperature change by addition weight of microwave susceptor.

The dissolution concentration of accumulated metal in aliquate of lake streams as regarding Pb heavy metal contamination is followed by equations of anion distribution, where n is kinetic order type (Eq. (4) and (5))

$$\frac{dc_{CO2}}{dt} = k_i c^{tin} dc \tag{4}$$

The dissolution concentration of accumulated metal in aliquate of sulfurous hot water streams as regarding Pb heavy metal contamination is followed by equation, where SO_4^{-2} sulphate concentration in effluent, $.f_i$ is concentration rate of sulphate

$$\frac{dc_{SO_4^{-2}}}{dt} = k_i c^{tin}.dc.f_i(SO_4^{-2})^{tin} \tag{5}$$

The dissolution concentration of accumulated metal in aliquate of limestone rocks dissolution by hot water streams in subground lakes with high CO2 gas dissolved streams as regarding Na metal carbonation is followed by equation, where HCO_3^{-2} bicarbonate concentration in effluent, Bicarbonate sulphate hot streams reduce aciditiy as below oxygen consumption Eqs. 6 and 7

$$\frac{dc_{Na}}{dt} = k_i c^{tin}.dc.f_i(HCO_3^{-2})^{tin} \tag{6}$$

The dissolution concentration of accumulated metal in aliquate of high fertilizer dissolution by wrong amount of fertilizer use in theagricultural fields discharged to streams as regarding carbonation is followed by equation, where HNO_3^{-2} nitrate concentration in effluent

$$\frac{dc_{Na}}{dt} = k_i c^{tin}.dc.f_i(HNO_3^{-2})^{tin} \tag{7}$$

The dissolution rates of heavy metals in acidic mine waters and sulfurous hot streams occurred in the region of Ilısu Dam, Güçlükonak, Şırnak and Batman Province saline waters in Siirt and Şırnak. The contamination of some accumulated carbonation contents of hot streams and soils in the region are given in **Table 1** [7].

Effluent, mg/l	Şırnak Coal Mine Pool	Şırnak, Hezil Stream	Güçlükonak Hot Stream	Batman Hot Stream	Şırnak Kasrik Laguun	Ilısu Dam Laguun1	Ilısu Dam Laguun2
Hg	8,11	4,71	12,3	14,11	4,71	4,71	4,71
Pb	10,58	14,53	23,2	12,58	11,53	5,7	5,2
Fe	40,33	70,62	59	93,3	56,2	60,62	67,62
K + Na	7,52	8,46	8,7	8,52	8,6	≥70	≥50
Cd	24,72	19,56	14,1	14,72	19,56	16	15
Mn	33,3	24,1	24,2	43,3	24,1	≤25	≤25
Cu	27,2	30,2	15,7	7,2	10,2	≤15	≤15
As	1,10	2,44	2,8	2,10	2,44	≤5	≤5
CO3	0,57	0,37	1,9	0,67	0,55	≤15	≤15
Soil, ppm							
Hg	34,11	48,71	52,3	54,11	40,71		
Pb	10,58	24,53	23,2	20,58	11,53		
Fe	4,33	7,62	5,9	9,33	5,62		
K + Na	74,52	81,46	81,7	84,52	88,6	≥70	≥50
Cd	24,72	9,56	10,1	4,72	19,56		
Mn	2,72	3,02	1,5	0,72	1,02	≤5	≤5
Cu	3,33	2,41	2,4	4,33	2,41	≤5	≤5
As	1,10	2,44	2,8	2,10	2,44		
CO3	0,57	0,37	1,9	0,67	0,55		

Table 1.
Şırnak and batman province reveals the potential carbonation scale and high saline bi carbonatewater source.

The use of water resources (water withdrawal and ordinance) and evaluation for development and community needs have been studied. However, the amount and quality of water that the eco-system will need is not addressed. Everything is built on the theme of "develop-supply-use". Parameters considered in the planning of water resources were population estimate, per capita water demand, fish farming, agricultural production, economic productivity level.

The agricultural irrigation, and human needs, the amount of water a healthy ecosystem will need, or actual regional water availability. The next step in traditional planning is to identify projects that will reduce the gap between estimated water supply and demand. In every scale, the planning action (region, basin, city) is used for the regular and healthy spatial development uses (housing, commerce, industry, recreation, other social) in the metropolitan cities which are especially migrating in our country and in medium size settlements Such as equipment) as directed by location decisions; It also determines the water demand of the city at the same time with its population and density of buildings and its quality and quantity of usage. While city plans shape the socio-economic and physical structure of the city, with the proposed land use, employment, population and density decisions, the city's daily water demand is also shaped. Therefore, any kind of urban development outside the plan creates an unhealthy environment that affects the quality of life of the city, as well as poses a serious threat to the water resources [6]. The quality of the water quality will be preserved, improved and monitored. Heavy

	SiO_2	Al_2O_2	Fe_2O_2	MgO	K_2O	CaO	TiO_2	LOI*
Kaolin (%)	47.85	37.60	0.83	0.17	0.97	0.57	0.2	11.27
Şırnak Asphaltite Char Shale	27. 54	7.70	10.83	2.17	1.97	10.5	1.74	5.47
Bentonite	50.45	17.80	6.83	12.17	4.97	3.57	0. 4	7.37
Marly Shale	17.85	11.60	0.83	5.17	3.97	20.57	0.4	5.27
Fly ash	27.8	13.60	17.83	4.17	2.97	10.7	1.4	16.27

**LOI: Loss on Ignition at 1000°*

Table 2.
Sorbent clay types for waste water treatment.

metal contamination hazard maps will be prepared and an early warning system will be established (**Table 2**).

3.1 Langmuir Crystallization model

For linear correlated distribution by depending on saturation limit and oxygen content of contaminant levels, the general common method can be given in **Table 1**.

The first order sorption concentration at three stage cycling counted but the equation below:

$$lnc^{Pb} = 1 + k_1t + f^t, 1ppm < x < 500ppm \tag{8}$$

3.2 Sludge carbonation by microwave radiation

The bentonite sample used in the study was obtained from Unye Madencilik from the Unye region of Tavkutlu mine. The bentonite sample was sieved and a small part of 45 μm was used for the operation. Bentonite samples were activated with 1 and 2 M HCl solutions for 2 h at 90° C using the Batch method (using 100 ml

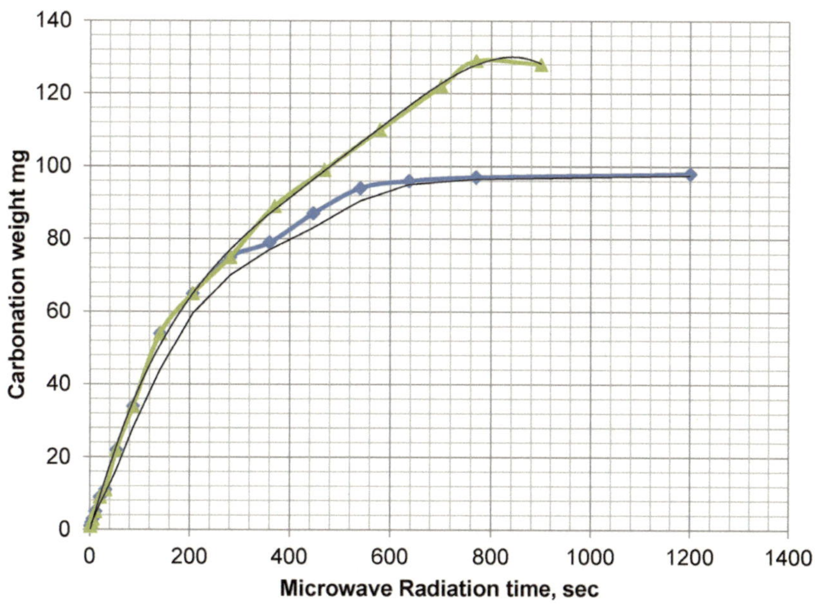

Figure 5.
Effect of caustic type on carbonation weight as jarosite and gibsite of gaseous CO2.

acid solution for 5 g sample). The acid-treated samples were washed with hot deionized water to remove Cl-ions and dried in room condition.

The microwave activated shale and fly ash used in the experimental work was provided from the district of Şırnak province. For the first time, Ünye region bentonite 0.1 M 100 ml CaCl2 solutions were mixed in the beaker at room temperature for 24 hours. and the filtrate was converted to the ion-exchange by applying the AgNO3 test. Acid/ clay suspensions were then prepared with bentonite, which was made to be ionized, to give H2SO4/clay ratios of 2 M, These were named - bentonite. These suspensions were dried at 150 oC for 3.5 hours. 50 ml of distilled water was added to the hot dry samples. The filtrate was filtered and dried at 80 oC. Finally, 5 ml of chlorite was added to the acid activated bentonite samples, each of which was 0.01 g, and the shale was retained by exposure to methanol vapor at 60° C for 4 hours. In addition, these samples were further dried at the same temperature for 1 hour to remove weak chlorite species. In this research, the caustic slurries microwave radiated cycled on the different bed sorbent types of double tube hot slurries radiated, cycled double washing sodium and calcium caustic content and bicarbonate precipitate samples of 10-20 gr were balanced. Precipitated particle size and weight distributions with fly ash sarnples are illustrated in **Figure 5**.

4. Results and discussion

The current use of absorbent bentonite and new areas of use increase in demand due to outflow. Especially absorbent claymarket, the cat and cat market A significant improvement in Americalt was. For absorbent clay depositsour wealthy country, too, to have a significant share ofthere is no reason why. This, which is gaining in Turkey, absorbant is limited to meet clay consumption [8–10].

This sorbent and carbonation caustic washing is done by compressing the CO2 to a dense phase as "supercritical". This supercritical phase is achieved by exposing the CO2 to temperatures higher than 32° C and pressure greater than 74 bars. The density of CO2 will increase with geo formation depth, until about 800 meters or greater, where the injection carried into a cap reservoir state [11–21].

Effective sorption in combustion processes depend on numerous factors including coal rank in carbonization, the volatile gaseous matter of coal such as presence of hydrogen, carbonyl gas and oxidation rate so stabilizing the desorbance, the settings of optimal diffusion conditions including structure defects (nitrogen, phosphorus, sulfur, etc.), temperature, oxygen content of coal, etc. and optimization of carbondioksit concentration ratios added the adsorption–desorption balance, the residence time and the spatial distribution of molecules in coal pores among other factors determining the efficiency of carbonization. as factors affecting the rate and extent of carbonization much dependent on the site activation, its desorption properties and its porosity. As discussed in the previous section, carbonization is a prerequisite step for oil generation from biomass wastes and coal [11, 12, 22, 23].

A major reason is that the retention time in fixed film processes is longer than in solid–gas processes. This allows more time to the carbonization far cracking to the desorbed persistent compounds. Furthermore, high rank coals allows an sufficient intimate contact between surface pores and gas atmosphere in the furnace due to more gas desorptions [13–16].

$$[(Na_2CO_3)n{\bullet}NaOH{\bullet}Na^+ + .n\ CO_2 = Clay/Ash + [(Na_2CO_3)n{\bullet}NaOH{\bullet}(NaCO_3)] \quad (9)$$

$$[(Na_2SO_3)2{\bullet}(Na)]^+ + n\ CO_2 = [(Na_2SO_3)n{\bullet}NaOH{\bullet}(NaCO_3)] \quad (10)$$

Potential storage areas are abandoned or in-production oil and natural gas fields, deep aquifers, soda mineral salt caves, methane-containing coal deposits and natural CO2 fields. **Figure 6** shows the distribution of oil and gas fields in Turkey. In **Figure 2**, the industrial facility and power with CO2 emissions determined within the scope of the project [24, 25].

Turkey discovered to have the largest natural CO2 field project, which is already the Field Raman EUR 7 billion Bm3 was produced with CO2 and used for landfill capacity. For CO2 storage projects, the properties of geological structures are important and almost all oil and gas fields characterization data are available. Since 1954. As a result of continued exploration work in more than 120 oil and gas fields have been found in Turkey [11, 22, 23]; However, the size of these fields is not very large and the largest field is West Raman, a heavy oil field. When the distribution of oil fields in Turkey is seen that almost all found in the vicinity of the South East Anatolia Region. Most of the natural gas fields are located in the Tigris Region. The target areas in this project are oil fields located in the South East Anatolia Region and close to a power plant or cement factory. Among these areas, Silopi Asphaltite Field was chosen due to its characteristics of having the least depth and highest pore volume, and having little or no cracks. The injected CO2 will be used as an EOR (enhanced oil recovery) tool to increase the amount of oil that can be produced before the storage phase. Cave of the limestone reservoir is few wells have been drilled in Raman Field so far, and production continues in 3 of these wells.

4.1 Pressurized sequential carbonation by Fly ash compost

The waste water washing provide the main support to the clean water production. The commercial successes in clay mud mentioned in washed bed and its sedimentation ability were described some of the emerging applications in lime use like clean water neutralization, aeration, [12–15]. The figures of statistical potential of washing control with different techniques in waste water cleaning are classified as seen in Figure 11 regarding cycling decantation time.

Some cost evaluations covering security of supply and environmental impacts, climate change evaluations, and technical and economic analysis, may be disused in cycling cost and activities. The jarosite type precipitations regarding sludge eşecropotantial of reduction occurred. There was also chloride hydrates formed by salt amount and pH alkalinity as given below:

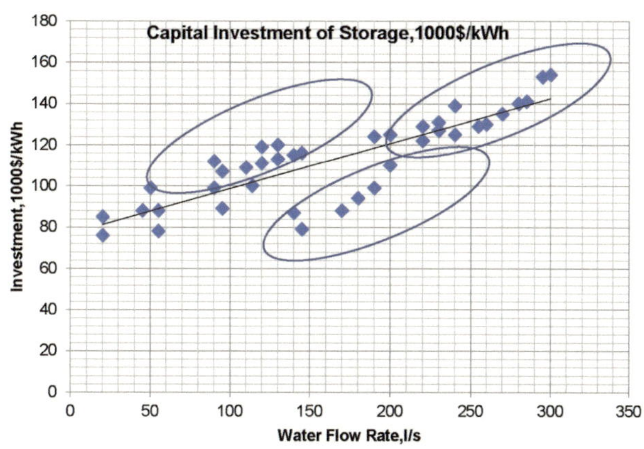

Figure 6.
The investment rates of compressed air/CO2 storage in vs capacity of combustion power plant.

$$\text{Clay} - Fe^{3+} + 4SO_2 + CO_2{}^- \leftrightarrow \text{Clay} - Fe(II) + 2S2O42 - +H_2O \rightarrow 2SO3^{2-} + S_2O_3{}^{2-} + 2H + HCO_3 \tag{11}$$

$$2CaO3(Fe_2{}^{3+}Al_{1.4}Mg_{0.6})Si_8O_{20}(OH)_4nH_2O + 2Na^+ + S2O4^{2-} + 2H_2O \leftrightarrow 2NaCa_{0.3}(Fe^{3+}Fe^{2+}Al_{1.4}Mg_{0.6})Si_8O_{20}(OH)_4nH2O + SO3^{2-} + 4H+ \tag{12}$$

Figures 7 and **8** showed The Waste sludge carbonation rates of Different Types of Sorbent Units cycled.

Initially, most of the toxin removal occurs through chemical adsorption of the toxins to the expanded clay where the combustion temperature was in the combustion phase below 750°C that lasts approximately 2–3 mins. The high surface area of fly ash particles and coal char was effective in the caustic slurries of 20 % weight rate Na and Ca caustic slurry with about 34 m2/g highly sufficient in order to react with gaseous CO2 10% of weights of fly ash were reacted in 10 bar caustic slurry of Na. Main reactive salt structure is sulphate in the caustic slurry covered widely pores associated active char.

A common industrial combustion to control the emissions pro combustion stage lime washing involves backwashing with air and hydrated lime water rinse. Process variables include the control backwash rate, surface wash rate/duration, time sequence and duration of backwash. Clean filtrate is pumped back into the bottom of the column during backwashing.

The required test and characterization procedures such as pH viscosity measurement, filtration loss and swelling index were applied to all bentonite concentrates obtained and then to products activated with 0.5% soda.

These sorbents need to be accurately mixed with combustion matter and to optimize the combustion process. Reliable models, based on the above results, need to be combustion chamber construction far the estimation of kinetic parameters for toxic stream control. Such toxic stream circulation models would aid in the

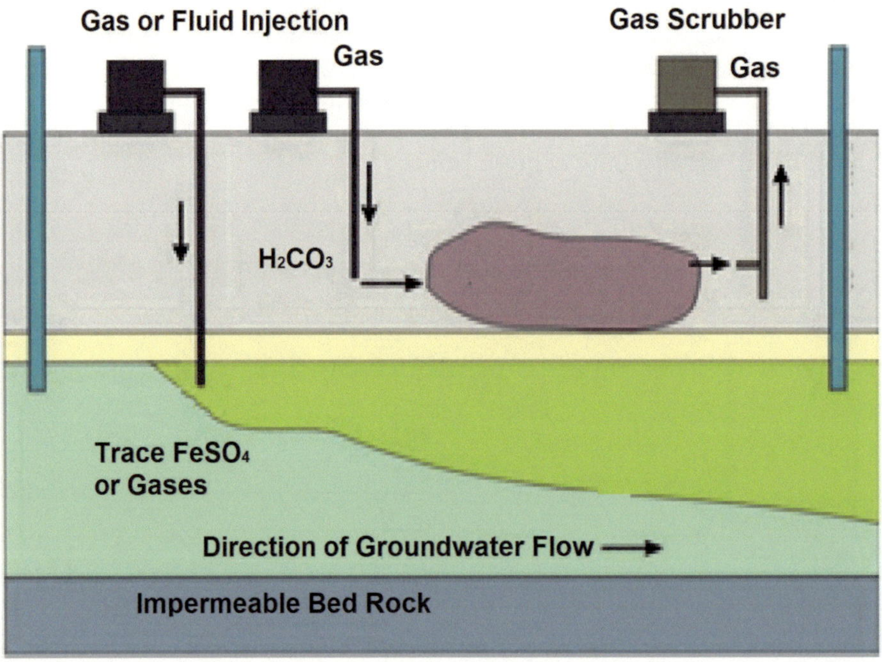

Figure 7.
The column carbonation sludge sequent in microwave activation as sorbent compost.

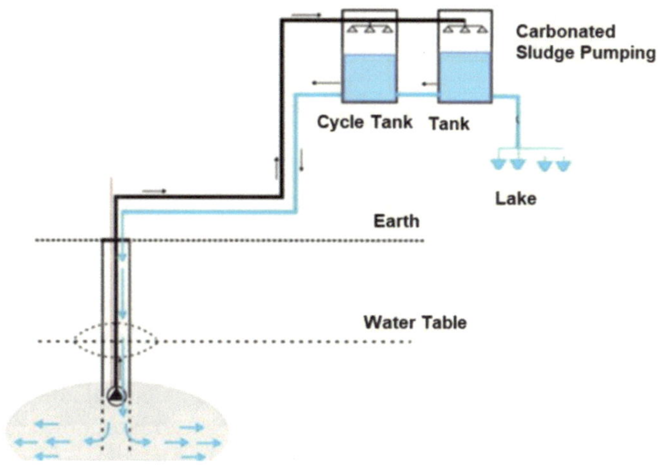

Figure 8.
The column carbonation sludge sequent in microwave activation as sorbent compost. The waste sludge carbonation.

microwave activated shale clay sorbent use in waste water treatment systems as shown in **Figure 5**.

The country needs the cleanest fuel to be produced providing the essential oils and gases. For this reason, acidic mine waters as heavy metal contamination to control fish farming were mixed with expanded clay at 1-2 mm size soaked with slurries of different alkali sorbents such as bentonite, shale fine, NaCl, CaCl2 and KCl were tested in the packed bed column washing and the test results were illustrated in **Figure 5**.

A carbonation bed column was used in compost sorption process was tested at a scale of 2–3 kg/h; collecting operational and design data to build an industrial installation. A technological diagram of the compost washing at three stage process developed unit was made. Activated shale destruction almost observed at third cycled end. Carbonate jarosite concentration change increased from 2nd cycle with performance of 30–40% and also simultaneous dilution of waste mud products by sedimented. it is necessary to optimize the cycling stages on metal circulation without the metal concentration change.

5. Conclusions

Reviews of the world shows that Turkey is responsible for only 1% of CO2 emissions. The establishment of the Carbon Market will ensure that the sectors and activities to be included in the evaluations will be determined.

- This study shows that, due to their low volume, known oil and gas reservoirs can only preserve the CO2 emissions of small industrial sites.

- In such a case, it is seen in this project that transportation of CO2 by tankers is more feasible.

- Silopi and Şırnak limestone a natural CO2 reservoir, can be considered as a suitable high-volume reservoir, currently with a suitable volume of 2 billion Sm3.

- The possibility of storage in deep saltwater aquifers should also be considered. A suitable pilot project will allow the parties to examine their eligibility.

- One of the most important elements of carbon capture and storage practices is incentives for CO2 storage.

The pH increased at washing was efficient in heavy metal sorption, the swelling index decreased, the loss of filtration increased negatively, and viscosity decreased by the addition of sodium.

In the obtained data, it was observed that sorption manner of bentonite has negatively effected by foreign ions in washing water for the activation especially total iron ion.

This result also indicated that the properties of the irrigation and fish farming water to be used during wet soil amendment of agricultural organic soil and lake muds with wet bentonites, on waste water treatment units which friendly mud should be controlled, otherwise the contamination after discharge would be harm human health, toxicology of animal and fish feed.

Abbreviations

Greek symbols

a	affinity parameter of the Langmuir isotherm (L mg^{-1})
b	stoichiometric constant defined by
B	reactant solid defined
Bi_m	Biot number for mass transfer
C_i	concentration of manganese in the bulk external phase of stage i (mg L^{-1})
C_O	feed concentration of manganese in the column (mg L^{-1})
D_{ef}	effective diffusion coefficient (m^2 s^{-1})
F	objective function
h	fixed bed height (m)
k_e	mass transfer coefficient in the bulk external phase (m s^{-1})
k_r	reaction rate constant for heterogeneous systems (m s^{-1})
N	number of stages
Q	volumetric flowrate (m^3 s^{-1})
q_i	concentration of immobilized manganese within the adsorbent particle at stage i (mg g^{-1})
q_m	theoretical maximum adsorption capacity of the Langmuir isotherm (mg g^{-1})
r	radial distance from the center of the particle, $0 < r < R_p$(m)
R	radius of column (m)
R_p	radius of adsorbent particle (m)
R^2	determination coefficient (–)
$r_{c,i}$	unreacted core radius at stage i (m)
t	time (s)
V_i	volume of stage i (L)
α	backmixing coefficient (–)
φ	column hold-up (–)
ρ	density of adsorbent particle (g m^{-3})
τ	mean residence time of fluid in the column (s)

Author details

Yıldırım İsmail Tosun
Mining Engineering Department, Engineering Faculty, Şırnak University, Şırnak, Turkey

*Address all correspondence to: yildirimismailtosun@gmail.com

References

[1] S. Ahamed, A. Hussam, A.K.M. Munir, 2009, Groundwater Arsenic Removal Technologies Based on Sorbents: Field Applications and Sustainability, Handbook of Water Purity and Quality, Academic Press, Amsterdam (2009) 379–417

[2] J.S. Ahn, C.M. Chon, H.S. Moon, K. W. Kim, 2003, Arsenic removal using steel manufacturing by-products as permeable reactive materials in mine tailing containment systems, Water Research, 37 (2003), pp. 2478–2488

[3] J.P. Allen, I.G. Torres, 1991, Physical separation techniques for contaminated sediment, N.N. Li (Ed.), Recent Developments in Separation Science, CRC Press, West Palm Beach, FL (1991)

[4] S.J. Allen, L.J. Whitten, M. Murray, O. Duggan, 1997, The adsorption of pollutants by peat, lignite and activated chars, Journal of Chemical Technology & Biotechnology, 68 (1997), pp. 442–452

[5] E. Álvarez-Ayuso, H.W. Nugteren, 2005, Purification of chromium(VI) finishing wastewaters using calcined and uncalcined Mg-Al-CO3-hydrotalcite, Water Research, 39 (2005), pp. 2535–25

[6] Volzone, C., Ortiga, J. 2009. Adsorption of gaseous SO2 and structural changes of montmorillonite. Applied Clay Science, 44, 251–254

[7] Gregg, S.J., Sing, K.S.W. 1982. Adsorption, surface area and porosity, Academic Press, London, 52 pp.

[8] Lopez-Gonzalez, J.D., Deitz, V.R. 1952. Surface changes in an original and activated bentonite. Journal of Research of the National Bureau of Standards, 48, 325–333.

[9] Marshall, C.E. 1935. Layer lattices and base-exchange clays. Zeıtschrift Fur Kristallographie, 91, 433–449.

[10] Murray, H.H. 1999. Applied clay mineralogy today and tomorrow. Clay Minerals, 34, 39–49.

[11] J.C. Abanades, M. Alonso, N. Rodríguez, B. González, G. Grasa, R. Murillo. Capturing CO2 from combustion flue gases with a carbonation calcination loop. Experimental results and process development. Energy Procedia **2009,** *1* (1) , 1147-1154. https://doi.org/10.1016/j.egypro.2009.01.151

[12] Vandersickel, A., Field, R. P., Chen, W , Mancini, N. D. and Mitsos, A. .2014, CaO-Based Energy and CO2 Storage System for the Flexibilization of an IGCC Plant with Carbon Capture. Industrial & Engineering Chemistry Research 2014, 53 (30) , 12032-12043. https://doi.org/10.1021/ie501475f

[13] **Tosun**, Y.İ, **2018,** *Microwave Caustic Leaching of Coal Slimes with Agricultural wastes - Production of Humate Pellets,* **CURRENT ACADEMIC STUDIES IN ENGINEERING SCIENCES-2018** Editors/ Editörler Prof. Dr. Serdar SALMAN- Prof. Dr. Rıdvan KARAPINAR Doç. Dr. Duygu KAVAK - Dr. Ali KILIÇER December 2018 / Cetinje-Montenegro, **ISBN • 978-9940-540-50-0, Ivpe** web: www.ivpe.me, Tel. +382 41 234 709, p689-700

[14] **Tosun**, Y.İ, **2016,** *Flue Gas CO_2 Sequestration by Turkish Coal Fly Ashes and Anatolian Geothermal Hot Waters* , ***Energy, Transportation and Global Warming***, (Ed.) Grammelis, Panagiotis, Springer International Publishing (Green Energy and Technology), Chapter 44: pp 605-616, ISSN 1865-3529 ISSNe 1865-3537, ISBN 978-3-319-30126-6, ISBNe 978-3-319-30127-3, DOI 10.1007/978-3-319-30127-3, Switzerland, May 2016.

[15] **Tosun**, Y.İ, **2018,** Thickener Water Neutralization by Mid-Bottom and Fly Ash of Thermal Power Plants and CO_2:

Organic Humate Mud of AMD Treatment for Remediation of Agricultural Fields, Chapter 8, pp141-170, ***Coal Fly Ash Beneficiation - Treatment of Acid Mine Drainage with Coal Fly Ash,*** Edited by Segun A. Akinyemi and Mugera W. Gitari, ISBN 978-953-51-3753-5

[16] **Tosun**, Y.İ. **2014, CO_2** Sequestratıon into shale beds - Şırnak coal mınes, 14th International Multidisciplinary Scientific GeoConference SGEM 2014, www.sgem.org, SGEM2014 Conference Proceedings, ISBN 978-619-7105-15-5 / ISSN 1314-2704, June 19-25, 2014, Book 4, Vol. 1, 101-108 pp, DOI: 10.5593/SGEM2014/B41/S17.014, Albena Resort, Bulgaria

[17] Ida, J.-I.; Lin, Y. S., 2003, Mechanism of High-Temperature CO2 Sorption on Lithium Zirconate. Environ. Sci. Technol. 2003, 37,1999-2004.

[18] Kato, M.; Yoshikawa, S.,2002, Nakagawa, K. Carbon dioxide absorption by lithium orthosilicate in a wide range of temperature and carbon dioxide concentrations. J. Mater. Sci., Lett. 2002, 21, 485-487.

[19] Hartman, M.; Coughlin, R. W. Reaction of Sulfur Dioxide with Limestone and the Influence of Pore Structure. Ind. Eng. Chem., Process Des. Dev. 1974, 13 (3), 248-253.

[20] White, C. M.; Strazisar, B. R.; Granite, E. J.; Hoffman, J.S.; Pennline, H. W. 2003,Separation and Capture of CO2 from Large Stationary Sources and Sequestration in Geological Formationss Coalbeds and Deep Saline Aquifers. *J. Air Waste Manage. Assoc. 53*, 645-715.

[21] Toraman OY, Depci T. Komurde Mikrodalga ile Onislem Uygulamalari. Madencilik, 2007;46:43-53

[22] Murray, H.H. 2000. Traditional and new applications for kaolin, smectite and palygorskite: a general overview, Applied Clay Science, 17, 207–221.

[23] M Broda, R. Pacciani, C. R. Müller, 2014, CO_2 Capture via Cyclic Calcination and Carbonation Reactions, *Porous Materials for Carbon Dioxide Capture Eds*An-H Lu, S Dai , Springer Berlin Heidelberg ,Print ISBN: 978-3-642-54645-7

[24] Jones DA, Lelyveld TP, Mavrofidis SD, Kingman SW, Miles NM. Microwave heating applications in environmental engineering—A review. Resources, Conservation and Recycling. 2002;34:75-90

[25] Kingman SW, Rowson NA. The effect of microwave radiation on the magnetic properties of minerals. Journal of Microwave Power and Electro-magnetic Energy. 2000;35:144-150

Chapter 4

COVID-19: A Learning Opportunity to Improve Environmental Sustainability

Syed Abdul Rehman Khan, Laeeq Razzak Janjua and Zhang Yu

Abstract

In just a few months, COVID-19 transformed from a dangerous regional health threat into a widespread global pandemic and economic disaster. Thus the world is expecting a great recession once again. The rapid spread of COVID-19 has had far-reaching consequences for people's daily lives in almost all parts of the world. Climate change and biodiversity depletion have now reached global boundaries; thus, human activity has surpassed Earth's capacities. Earth capacities can be explained in terms of extreme climate change. This chapter is intended to investigate the link between the outbreak of Covid-19 and its effect on environmental and society. The discussion reveals that environmental pollution is minimized as a result of global lockdown. Furthermore, our review also shows that in terms of environment, Covid-19 provide an opportunity to transform our polluted economy toward the green economy through adoption of renewable energy sources and green practices in our businesses.

Keywords: COVID-19, air pollution, climate change, carbon carpets

1. Introduction

The estimated age of this world is approximate (13.8 billion years). In all these years, there are some incidences experienced by a human, which reshaped the entire behavior of human. World war I, II, cold war era and more recently the incident of 9/II all these incidences in world history suddenly changed the power centers, the world moved from unipolar to bipolar or even toward multipolar world order. Furthermore, all these event or incidence dramatically changed the over economic systems, way of governing nations and even changed our literature as well. Even all these incidences changed the lifestyle of the entire world, similarly, the ongoing pandemic crises of covid-19 will have a permanent and everlasting impact on our lives [1, 2].

It is generally believe that once pandemic crises will over; the world will be not the same as it was before. Technically, due to hyper-growth, presence of high tech inventions, innovation and more especially the extensive use to the internet already made this world liquidness thus we are unable to identify where it will be stopped or freeze. On the other hand, nowadays, this pandemic situation just created the new standard among us; it is also true; thus, it locked down the entire world in one corner. At the

Year	Pandemic crises	No. of deaths
165 AD	Antonine Plague	5,000,000
541 AD	Justinianic Plague	100,000,000
1334	Black Death Plague	25,000,000
1576	Cocoliztli Epidemics	15,000,000
1817	The first Cholera Pandemic	150,000
1855	The third Plague	15,000,000
1889	Russin Flu	360,000
1919	Spanish flu	50,000,000
1957	Asian Flu	1,100,000
2004	SARS	813
2014	EBOLA	11,315
2020	Covid-19 (until 08-01-2020)	712,126

Table 1.
Death toll in pandemic crises in history (source: Author working sheet).

moment, due to this pandemic, the circle of growth has become stiff. Thousands of airlines are not functioning, and the majority of the tourist places are closed, furthermore there not any educational activities are going on in most effected countries of covid-19. Similarly; on the other hand, this world not facing the pandemic situation the first time. History reveals that there were numerous pandemic crises experienced by this world. Due to all these pandemic crises, millions of people lost their lives. **Table 1**, indicates major pandemic crises experienced by the world in history.

In the recent era, especially at the beginning of the 21st century, two more pre-mature pandemic crises hit numerous countries of the world. Avian influenza, which is known as bird flow virus, infected hundreds of people from 2003 to 2011 in almost 60 countries of the world. According to the World Health Organization estimation, more than 1500 people died due to this influenza. Furthermore, in 2009 swine flu (H1N1) spread over more than 100 countries of the world, and approximately 550,000 people lost their lives due to this (Centers for Disease Control and Preventions). On the other hand, from 2013 to 2016, the Ebola virus hit western African countries, and more than 11,000 people died. Similarly, there are numerous diseases which cause millions of death worldwide every year. Due to HIV (AIDS) 32 million people died, whereas due to malaria, just in 2018, approximately 400,000 people died (World Health Organization).

2. Climate change: why it matter?

Before discussing actual environmental issues, it essential to describe the phenomenon of climate change. United Nations Framework Convention on Climate Change defines it as "a change of climate which is attributed directly or indirectly to human activity that alters the composition of the global atmosphere and which is in addition to natural climate variability observed over comparable periods" [3]. Similarly, on the other hand, Rahman, [4] define it as 'our climate is changing, largely due to the observed increases in human-produced greenhouse gases. Greenhouse gases absorb heat from the sun in the atmosphere and reduce the amount of heat escaping into space. This extra heat is the primary cause of observed changes in the climate system over the 20th century. However one question arises here, why environment or typical climate change matter for us; as it represent our common future and that particular phrase of common future is defined by United Nation's Brundtland Report 1987 as 'sustainable development'. Leggett and Carter [5] refer definition of sustainable development, which is described by united nation conference on Sustainable development; Sustainable development meets the needs of the present without compromising the ability of future generations to meet their own needs. Seem like the guiding principle for long-term global development; sustainable development consists of three pillars: economic development, social development, and environmental protection.

The adverse effect of climate change is exceptionally massive; somehow, it matters of survival of humankind on earth. Sustainability can be achieved in a balanced way only if we care about the world. Among three pillars of sustainable development, which are environment, economic and social; environment considers as one of the most significant pillar of sustainable development. According to the UN estimations, for human survival till now 1.6 planet resources we already used and while observing the acceleration by which we (humans) utilize natural resources two piles of the earth will be required by 2030 (Global Footprint Network). As Foley et al. [6] mentions, in his ecological overshoot, he is concerned about the conversion of resources into waste faster and vice versa. The most noticeable effects of overshooting or overutilization of natural resources can be observed through diminishing forest covers, collapsing fisheries, rapid fall in water level under the earth crust, carbon dioxide emissions which all are creating global climate change. Global warming affects all species which are living on the earth. According to living planet report 2014, the population of living species has declined by 52% from 1970 (World Wildlife Fund). The idea of Sustainable Development Goals become more critical among intellectual, economist, and scientist due to the growing urgency of sustainable development for the entire world, as climate change and medical issues are those common challenges.

Table 2 illustrates some of the adverse features of climate change are massively noticed in recent decades. Due to industrialization and foreign direct investment outflow from developed countries to developing countries where environmental laws are not so much strict bad affect atmosphere. As massive harmful gases in the shape of Carbon dioxide, Sulfur dioxide and nitrogen dioxide are added in the atmosphere, which eventually decreases oxygen and harms our environment which

Climate changer features	Why it matters?
Green House Gases Concentration	Emission of Green House Gases thorough industrialization, traveling etc. is increasing the GHG concentration in the atmosphere.
Ozone Layer depletion	A slow, steady decline of about 4% per decade in the total volume of ozone in Earth's stratosphere (the ozone layer) since the late 1970s
Melting of Ice	Greenland lost 150 km3 to 250 km3 (36 mi3 to 60 mi3) of ice per year between 2002 and 2006 and Antarctica lost about 152 km3 (36 mi3) of ice between 2002 and 2005
Rise in Sea level	Global sea level rose about 17 cm (6,7 in) in the last century
Temperature rise in sea water	With the top 700 m (about 2300 ft) of ocean showing warming of 0,16□□C since 1969 due to absorbed increased heat of the Earth
Unwelcome rains and storms	World is experience every year unwell come rains and floods and thousands of people lost their life's.

Table 2.
Adverse affect of climate change (source: Author working sheet).

surrounds us. Similarly, one of the most destructive effects of all these harmful gases in our environment and atmosphere is depletion of the ozone layer which protect humans and surface of the earth from the adverse impact of sun rays. The concentration of harmful gases in the atmosphere increases the average temperature of the world, which negatively affect the glaciers and phenomenon of melting ice increases very rapidly. Melting of ice in summer season induces floods in our river, which causes loss of human life and another material loses as well. However, massive waves in the sea and similarly melting ice slabs directly increase sea level, which influences the life of people living on the seashores. The **Figure 1** indicates the average increase in world temperature recorded in the last 40 year. Thus we can see that average temperature indicating an upward trend, which is an indeed alarming situation for everyone.

The world-renowned, global climate change activist, Greta Thunberg Said, "And Yes I know that we need a system change rather than individual change, but you cannot have one without the other. If we explore the history, all the big changes in society have been started by the people at the grassroots level. No system change can come without pressure from the large group of individual'. Thus, it is right the entire world needs collective efforts to face all the challenges to arise by climate change. Moreover, the highest responsibility is on the shoulder of the business community, policy markers, world leaders and politicians. The business community employs by using the labor from the factor of production; moreover, they also put capital along with the utilization of land resources which are also the factor of production. However, unsustainable utilization of land resources creates massive climate change threats. Every year tons of plastic waste is buried without recycling under the surface of the earth and dump in oceans as well.

Similarly, in order to fight against hunger and our food necessities, our farmer's working day and night on land resources and grow crops and planted fruits trees. However, by using heavy pesticides and fertilizer we can double the quantity of wheat in less time but on the other hand, one of the significant adverse effects of pesticides are the death of eco-friendly insects. Similarly, extensive use of fertilizer is finishing the natural tendency of soil to grow crops.

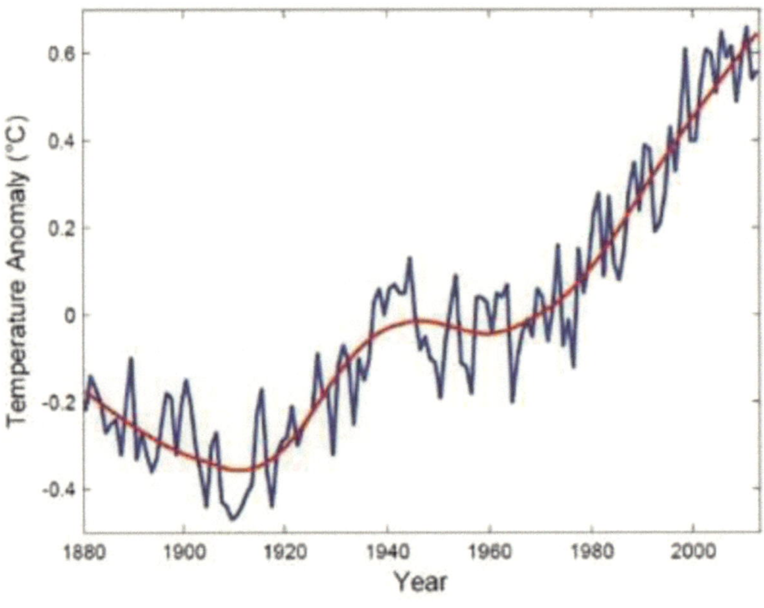

Figure 1.
Average temperature trend observed by NASA since 1980. Source: Rahmstorf et al. [7].

Every next coming day increases the world population in such manner that, every consecutive day noted as the highest population ever recorded in the world. Similarly, our rural areas are finishing very rapidly. On the other hand, urban concentration is increasing, which cause a heavy abnormal burden on cities. One of the most major adverse effects can be observed in developing countries and more specifically in unplanned cities as the majority of unplanned cities in the world are expanding in such manner that majority of the people living there without basic human needs such as access of freshwater, sewerage system and some of the places even with the access of electricity. The UN Framework Convention on Climate Change [8] in their report predicted that billions of people in developing countries would face shortages of water, food and more significant risks to health. Thus it requires immediate action in a sustainable way along with effective planning and then assessments. The issue which arises due to climate change will not be solved suddenly; in fact, it is a long term process which requires proper planning and contribution of every stakeholder and individual. As we already discussed earlier, it is not an issue of one individual, single community or a country's problem. Thus its global matter, therefore; requires worldwide attention and contribution.

Similarly, as **Figure 2** indicates the sunlight rays pass through the ozone layer, which eventually increases the temperature. Achieving sustainability requires collective action by the entire stakeholder of society; in fact, it requires action from the farmer working in the farm and earning for his family as well as a high corporate executive. One of the significant issues now a day's faced by our society is income inequalities or unequal distribution of wealth, which divide our societies into different classes. One of the most destructive factors which arise in our society is that the rich become richer and the poor become poorer. Similarly, income inequalities can be observed in developed countries and as well as in developing countries. In developing countries usually, rich people have access toward the higher political level.

One of the most challenging situations can be observed in developing countries in which wealthy class, provide access to the foreigner investor in their country and due to weak local legal system and absence of environmental laws and regulations, heavily polluting industries is formed in host countries which adversely affect the

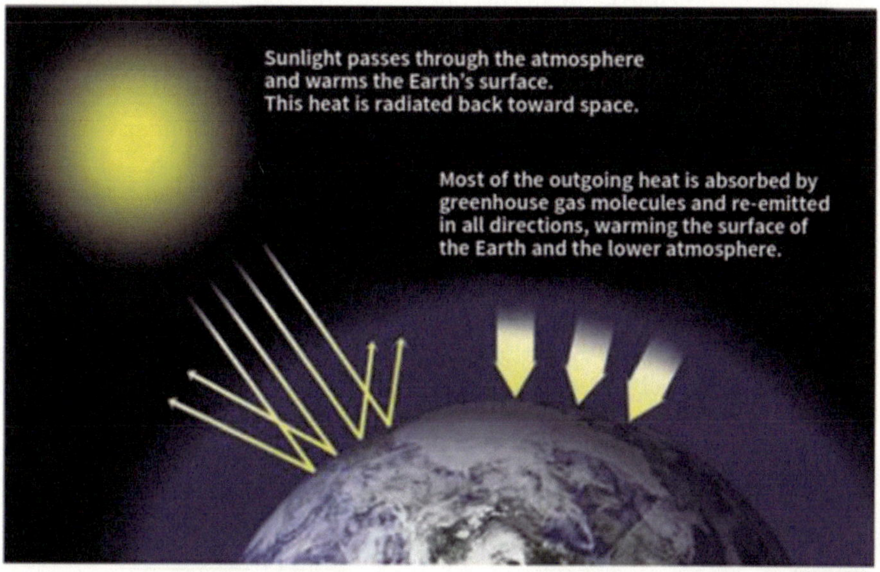

Figure 2.
Ozone layer and our world. Source: Global Climate Change, [9].

atmosphere. However, recently it had also been noticed that polluted industries also induce health issue for the local communities, including lungs cancer and asthma. The more adverse effect can be observed in developing countries in the shape of massive life losses where medical facilities are not so much advance and widely available. Apparently, in such situation, it is the responsibility of all the stakeholders to identity such industries which create an environmental hazard as well as, higher leadership should also take strict action to fight against global warming issues [10]. On the other hand, it has also been observed that in developing countries the industrial waste is directly dumped into the rivers and sea without any further recycling [11]. Therefore life undersea and river profoundly affected as well. Similarly, the extensive use of plastics also harms our surroundings.

3. Nexus between Covid-19 and environment

Air pollution is one of the most considerable catalyst factors which is harming our environment. Air Pollution defined as release of poisonous gases, which are emitted by factories and transportation. As due to the pandemic crises, from beginning of the year most of the countries of the world were in lockdown situation. Due to the fact, lock down situation ultimately affects every sector of the world. The overall world output dramatically decreased due to the lockdown and millions of the people lost their jobs. **Figure 3**, indicate impact of Covid-19 on different economic activities around the world.

Similarly an empirical study conducted by Barro et al. [12] reveal due pandemic crises on average 2.1% death rate could cause decline of averagely 6% of world GDP and 8% decline in private consumption. Furthermore, in terms of macroeconomic impact, Coibion et al. [13] empirically analysis pandemic impact using survey on household in the US. They conclude that, due to pandemic crises consumption and employment decreased where as inflation and economic uncertainty increased.

It is true to argue that, due to the massive lockdown in the world overall production and consumption decreased. Lockdown indicate downward trend of industrial output and due to 'stay safe stay health' less transportation vehicles were used by people which directly impact on the overall air pollution of the world [14].

In terms of empirical studies, Brodeur et al. [15] mention in their work 'safer-at-home' policies indicatively decreased air pollution across US counties. Therefore we can argue that one the key factor which contribute downward trend of air pollution is transportation vehicles thus consumption of gasoline as fuel. Similarly, Cicala et al. [16] identified in USA the average sales of gasoline from 2007 to 2019 average recorded as 8000 to 9000 barrels per day, however due to the lockdown and less mobility, the average sales drop at 5000 barrels per day until 12th week of 2020 (**Figure 4**).

On the hand, it is also true beside industrial sector and road transportation the large consumption sector of fuel is airline industry, although due the lockdown situation massive airlines of the world grounded there air craft's and international mobility stopped as well. Indeed, that particular factor also decreases overall air pollution.

It has be also observed that, air travel thus airline industry account for about 2.5% of global CO2 emission; furthermore it is predicted that until 2050 airline industry could take up a quarter of the world's carbon budget and it will cause for 1.5 degree Celsius world temperature [17]. **Figure 5**; indicate first quarter international flights and transportation activity around the world.

In terms of china, Almond et al. [19] investigate nexus between air pollution and gases emission during covid-19 crises. They conclude that during the crises NO2

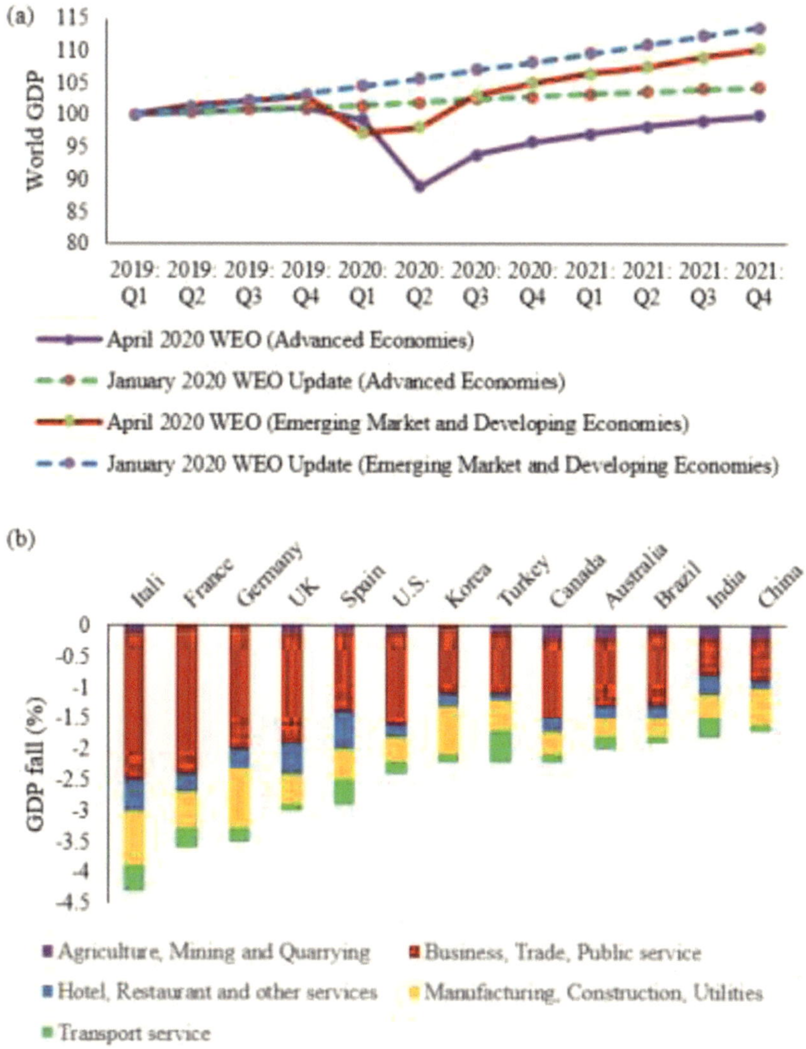

Figure 3.
Quarterly word GDP.

emission decline whereas SO2 did not fall. Furthermore they also mention that, in overall China (PM2.5) emission decreased by 22%. The major cause of reduction in air pollution is due to less personal vehicle usage which is heavily responsible for NO2 emission. Similarly, in case of china another study conducted by He et al. [20] also concluded same results, due to the lockdown in china PM2.5 concentration decline by 25%. That particular phenomenon were recorded in more industrialized cities, therefore it is true to argue that the fall in industrial production is positively associated with reduction in air pollution. One more study conducted by Yao et al. [21], mention that higher humidity causing spread of the Covid-19, and they further illustrate that high level of NO2 emission, playing role as catalyst agent and causing more Covid-19 spread with in the china.

After the lockdown, the clean environment has been observed all around the world, not only in high industrialized countries, more specifically in European countries due to the lockdown situation, and the concentration of NO2 emission in the air rapidly decreased. **Figure 6**, illustrate the decline of NO2 emission in most of the European countries.

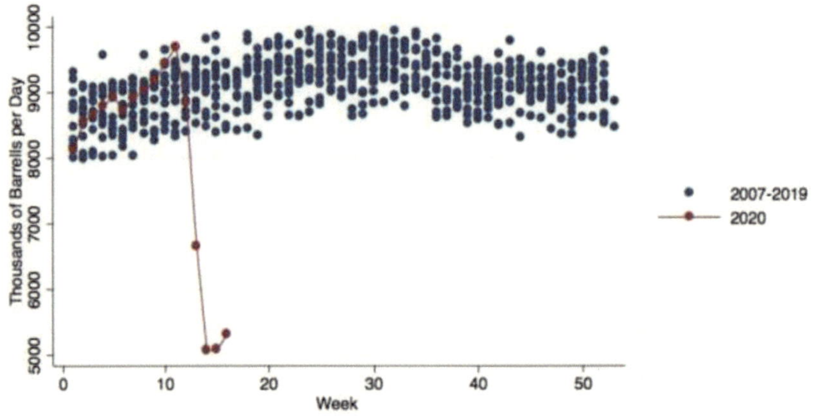

Figure 4.
Sales of gasoline in USA from 2007 to 2020 (until 12th week). Source: Cicala et al. [16].

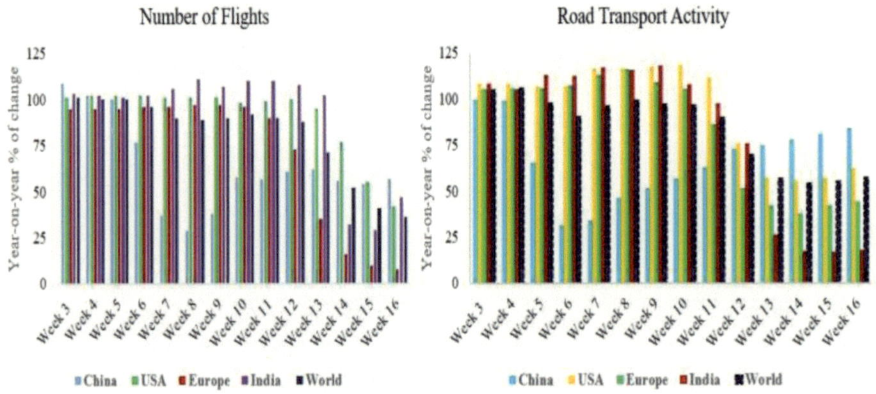

Figure 5.
International flight and transportation activity around the world. Source: International Energy Agency [18].

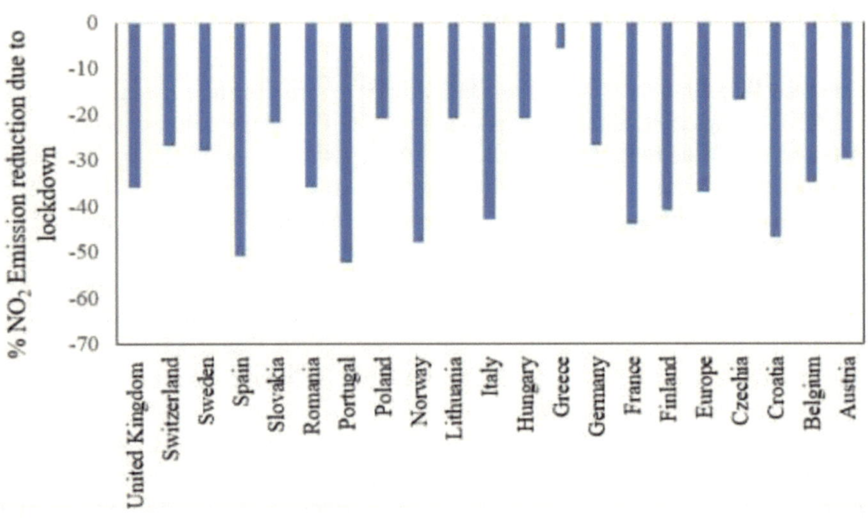

Figure 6.
Percentage change of NO2 emission in European countries. Source: Myllyvirta and Thieriot [22].

Similarly not only European countries have to experience clean air, on the other hand, a country like India, which is the second-highest populated country of the world, due to the lockdown situation the trend of air pollution also declined in India [23]. It is also true that, due to lockdown in massive countries of the world, the overall pollution decreases rapidly; thus, environment restores as it was 50 year ago. Not only, lockdown reduce air pollution; eventually, it also impacts on water pollution. In Europe, the death toll of covid-19 record massively high in Italy, therefore during the lockdown in Italy tourist arrival become zero almost, which impact on water quality in the canals of Venice. After couples of decades, the resident of Venice city has noticed clean water and marine life in the channels again [2]. The significant declines which has also been observed in the pattern of the overall consumption of electricity. As many of the people working from home, this directly decreases electricity consumption. Furthermore, the countries which heavily depend on electricity production via coal have experienced a decline in CO2 emission as well. It is right to argue that covid-19 ultimately reduced the consumption of electricity and fuel consumption for producing electricity as well [24]. Below **Figure 7** indicate, the overall pattern of NO2 emission in Europe during 2019 and 2020, thus after the lockdown situation.

The positive change and transformation in environment is generally due to the lockdown situation in terms of clean air and clean water channels, will not be

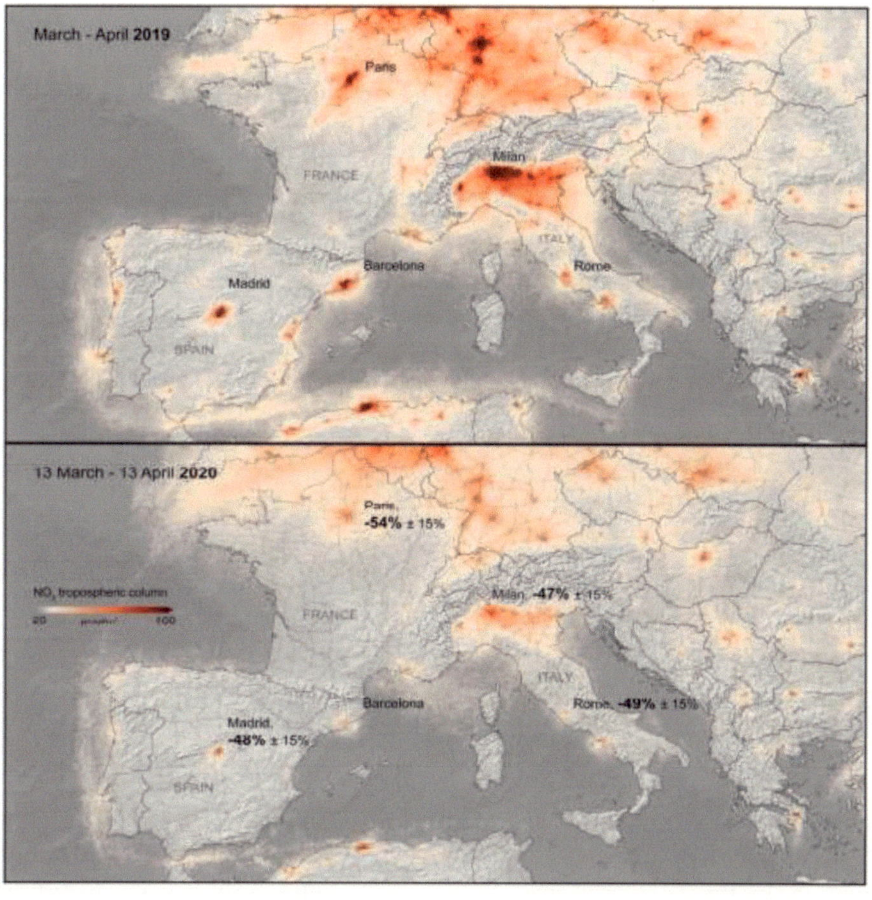

Figure 7.
NO2 emission in Europe comparative analysis between March April 2019–2020. Source: European Space Agency, [25].

long-lasting and it is expected that our environment once again will harm as the world will experience normal cycle of life; thus we will again experience massive air and other pollution [26]. Massive corporate allow remote work for their employees. Remote work will enable people less mobility, thus, which decrease the usability of the vehicle. However at the moment it is not responsibility of single person or a country to maintain that transformation, thus we should promote a green ideology, which will ultimately reduce air pollution and on the other hand green concept will improve countries' social, economic and environmental performance. Furthermore, we should also focus on renewable energy in overall logistics operations, as it ultimately improves environmental performance by reducing emission whereas environmental performance is negatively correlated with public health expenditure as explained in literature [27].

4. Conclusion and recommendations

The aim of this chapter is to discuss the association between Covid-19's outbreak and its environmental effects on society. The debate indicates that the global lockout eliminates environmental emissions. As a result of the global shutdown, we have induced economic downturn across the world, but on the other hand, in a better climate, we are gaining a transformed world. There is no doubt that once the world engine start-up after Covid-19, pollution and waste will started to piling-up in our surrounding, which will not damaging to the health but also environmental sustainability. The implementation of different step-by-step strategies demonstrates that the environment will gradually shift from polluted world to a green one. In order to maintain environmental sustainability, policymakers and governmental bodies should develop a road-map to protect environmental sustainability.

Following are some suggestions, which may help in the initial phase of planning.

- Corporate sector should adopt green and eco-friendly practices in their businesses to improve environmental sustainability.

- Governmental bodies should formulate strict eco-friendly policies and provide awareness to the corporate leaders through different workshops.

- Regulatory bodies may encourage renewable energy and green projects through providing subsides and tax-exemptions to the enterprises.

- Governmental bodies embossed heavy penalties on the polluted firms, which will not only create pressure on firms to adopt sustainable practices but also motivate to the eco-friendly firms.

- Regulatory authorities may increase the fossil fuel and non-green energy prices, which will ultimately create pressure on firms to adopt renewable energy in their operations.

Furthermore, In order to achieve industrial productivity for green growth, the reliance upon energy use of fossil fuels should be replaced with renewable energy sources. As for the sustainable development of the country, the efficient use of natural resources, including oil rentals, ores and metal exports is imperative. On the other hand, Green finance will offer an incentive to fund sustainable items that lead to the digital mobilization of renewable energy and industrial growth initiatives.

Furthermore, economic policies will help to achieve a smooth growth agenda through capital market promotion, technological advances, and the provision of environmental subsidies, to reduce the environmental adverse externalities of the production and consumption sectors. In the last, Public-private cooperation on the resource market is necessary in order to support sustainable economic development, which makes it incredibly necessary to provide insurance and business services such that green development is encouraged.

Acknowledgements

This work supported by the China Postdoctoral Science Foundation (No. 2019 M660700), and the Beijing Key Laboratory of Megaregions Sustainable Development Modeling, Capital University of Economics and Business (No. MCR2019QN09).

Author details

Syed Abdul Rehman Khan[1,2]*, Laeeq Razzak Janjua[3] and Zhang Yu[4]

1 School of Economics and Management, Tsinghua University, Beijing, China

2 School of Management, Xuzhou University of Technology, Xuzhou, China

3 Poznan University of Economics and Business, Poland

4 School of Economics and Management, Chang'an University, Xi'an, China

*Address all correspondence to: sarehman_cscp@yahoo.com

References

[1] Andree, B. P. J. (2020). Incidence of COVID-19 and Connections with Air Pollution Exposure: Evidence from the Netherlands.

[2] Saadat, S., Rawtani, D., & Hussain, C. M. (2020). Environmental perspective of COVID-19. *Science of The Total Environment*, 138870.

[3] UNFCCC, D. (2011). 1/CP. 17 in Report of the Conference of the Parties on its Seventeenth Session. In United Nations Framework Convention on Climate Change.

[4] Rahman, M. I. U. (2013). Climate change: A theoretical review. *Interdisciplinary Description of Complex Systems: INDECS*, *11*(1), 1-13

[5] Leggett, J. A., & Carter, N. T. (2012, June). Rio+ 20: The United Nations Conference on Sustainable Development, June 2012. Library of Congress, Congressional Research Service.

[6] Foley, H., Bogue, J., & Onakuse, S. (2016). New conceptual framework for sustainability. Irish Studies in International Affairs, 27, 145-163.

[7] Rahmstorf, S., Foster, G., & Cahill, N. (2017). Global temperature evolution: recent trends and some pitfalls. *Environmental Research Letters*, *12*(5), 054001.

[8] Intergovernmental Panel on Climate Change (IPCC). (2014). AR5 synthesis report: Climate change 2014.

[9] Global Climate Change, (2020) https://climate.nasa.gov/ [Access on 31-07-2020]

[10] Zambrano-Monserrate, M.A., Ruano, M.A., Sanchez-Alcalde, L (2020) Indirect effects of COVID-19 on the environment, *Science of The Total Environment*, 728, 138813

[11] Khan, S. A. R., Sharif, A., Golpîra, H., & Kumar, A. (2019). A green ideology in Asian emerging economies: From environmental policy and sustainable development. *Sustainable Development*, 27(6), 1063-1075

[12] Barro, R. J., Ursúa, J. F., & Weng, J. (2020). *The coronavirus and the great influenza pandemic: Lessons from the "spanish flu" for the coronavirus's potential effects on mortality and economic activity* (No. w26866). National Bureau of Economic Research.

[13] Coibion, O., Gorodnichenko, Y., & Weber, M. (2020). *The cost of the covid-19 crisis: Lockdowns, macroeconomic expectations, and consumer spending* (No. w27141). National Bureau of Economic Research. *Complex Systems: INDECS, 11*(1), 1-13.

[14] Eroğlu, H. (2020) Effects of Covid-19 outbreak on environment and renewable energy sector. *Environ Dev Sustain*. https://doi.org/10.1007/s10668-020-00837-4

[15] Brodeur, A., Gray, D. M., Islam, A., & Bhuiyan, S. (2020). A Literature Review of the Economics of COVID-19.

[16] Cicala, S., Holland, S. P., Mansur, E. T., Muller, N. Z., & Yates, A. J. (2020). *Expected Health Effects of Reduced Air Pollution from COVID-19 Social Distancing* (No. w27135). National Bureau of Economic Research.

[17] Tabuchi, H. (2019). 'Worse Than Anyone Expected': Air Travel Emissions Vastly Outpace Predictions. *The New York Times*.

[18] International Energy Agency, (2020) https://www.iea.org/ [Access on 25-07-2020]

[19] Almond, D., Du, X., & Zhang, S. (2020). *Did COVID-19 Improve Air*

Quality Near Hubei? (No. w27086). National Bureau of Economic Research.

[20] He, X., Lau, E. H., Wu, P., Deng, X., Wang, J., Hao, X., ... & Mo, X. (2020). Temporal dynamics in viral shedding and transmissibility of COVID-19. Nature medicine, 26(5), 672-675.

[21] Yao, Y., Pan, J., Liu, Z., Meng, X., Wang, W., Kan, H., & Wang, W. (2020). Ambient nitrogen dioxide pollution and spread ability of COVID-19 in Chinese cities. *medRxiv*.

[22] Myllyvirta, L., & Thieriot, H. (2020). 11,000 air pollution-related deaths avoided in Europe as coal, oil consumption plummet. Available in: *https://energyandcleanair. org/wp/wp-content/uploads/2020/04/CREA-Europe-COVID-impacts. pdf. (Accessed May 2020).*

[23] Mitra, S. S., Costa, M., Joseph, K., & Notts, R. M. THE BOON IN THE BANE: INDIA GOES CLEAN-GREEN AMIDST CORONAVIRUS LOCKDOWN. *EPRA International Journal of Multidisciplinary Research (IJMR)*, 25, 233

[24] Lau, H., Khosrawipour, V., Kocbach, P., Mikolajczyk, A., Schubert, J., Bania, J., & Khosrawipour, T. (2020). The positive impact of lockdown in Wuhan on containing the COVID-19 outbreak in China. *Journal of travel medicine*, 27(3), taaa037.

[25] European Space Agency (2020) https://www.esa.int/ [Access on 01-08-2020]

[26] McCloskey, B., & Heymann, D. L. (2020). SARS to novel coronavirus–old lessons and new lessons. *Epidemiology & Infection, 148*.

[27] Khan, S. A. R., Zhang, Y., Kumar, A., Zavadskas, E., & Streimikiene, D. (2020). Measuring the impact of renewable energy, public health expenditure, logistics, and environmental performance on sustainable economic growth. *Sustainable Development*.

Chapter 5

Impact of Hybrid-Enabling Technology on Bertrand-Nash Equilibrium Subject to Energy Sources

Ryle S. Perera

Abstract

In this chapter, we quantify an optimal level of subsidy for the sharing of hybrid-enabling technology innovation in an energy market while examining its Bertrand-Nash equilibrium. We formulate this as a Stochastic Differential Game (SDG) and analyze the stability of the Stuckenberg, Nash and cooperative equilibria via a feedback control strategy. We then adopt limit expectation and variance of the improvement degree to identify the influence of the external environment on the decision maker. We show that the game depends on its parameters and the equilibria chosen. Ultimately, our use of short-run price competition characterized by strategic supplies for renewable and fossil resources provides a more robust model than that presented by Bertrand-Edgworth with endogenous capacity. As a result, we highlight that R&D investments in hybrid-enabling technology can ensure immediate reliability and affordability within energy production and implementation of policy instruments.

Keywords: Bertrand duopoly game, cooperative game, hybrid-enabling technology, Nash non-cooperative game, Stackelberg game, stochastic differential game

1. Introduction

In recent years, many researchers have developed models to discuss the importance of lowering carbon emissions and its potential impact on society by examining economic growth, international trade, and health benefits. Khan et al. [1] examined the relationship between green logistics indices, economic, environmental, and social factors through the perspective of Asian emerging economies. By adopting a Fully Modified OLS (FMOLS) Model and Dynamic OLS (DOLS) they claimed that logistics operations, particularly the efficiency of customs clearance processes, quality of logistics services and trade and transport-related infrastructure are positively and significantly correlated with per capita income, manufacturing value added and trade openness, whereas greater logistics operations are negatively associated with social and environmental problems including, climate change, global warming, carbon emissions, and poisoning atmosphere. Khan et al. [2] examined the potential relationship between public health expenditures, logistics performance indices, renewable energy, and ecological sustainability in members of the Association of Southeast Asian Nations by applying the structural equation modeling approach. They showed that the use of renewable energy in logistics operations will improve environmental and economic performance to reduce emissions, whereas environmental performance is negatively correlated with public health expenditures, indicating that greater environmental sustainability can improve human health and economic growth. In [3], economic growth and environmental sustainability in the South Asian Association for Regional Cooperation using the data from the South Asian Association for Regional Cooperation (SAARC) member countries from 2005 to 2017 was examined. Adopting the panel autoregressive distributed lag technique to examine the hypotheses, they find that environmental sustainability is strongly and positively associated with national scale-level green practices, including renewable energy, regulatory pressure, eco-friendly policies, and the sustainable use of natural resources. In [4], the consumption of renewable energy with international trade and environmental quality in Nordic countries from 2001 to 2018 is investigated. Their findings concluded that renewable energy is strongly and positively associated with international trade in Nordic countries. Furthermore, [5] adopted multi-criteria-decision-making techniques to examine barriers in the sustainable supply chain management (SSCM) when firms are facing heavy pressure to adopt green practices in their supply chain (SC) operations to achieve better socio-environmental sustainability.

Around the world governments, businesses and individuals have committed to reducing carbon emission. As a result, the energy economy is highly exposed to these processes. As industries push for renewable energies, technology will need to step in to ensure reliability of the power supply. Therefore, there remains a need for exploiting the role of hybrid technology, its dynamics, limitations on the reduction of pollution levels and policy implementation within the wider carbon emissions debate. This is due to the vital role hybrid technology plays in energy production processes and the ability for the energy system to offer a better energy security. Development of such lower carbon emission policies has potential benefits to the environment and ecological sustainability to those economies. However, within many of these environmental policy models, technology is incorporated as an exogenous variable and limited attention is given to endogenous technology, other technological breakthroughs, potential government subsidies or collaborative innovations to integrate low carbon technology in environmental economics. Such interventions will promote the renewable energy sector to use natural resources and undertake public-private partnership investments to minimize dependence on fossil fuel derived energy.

To investigate the effects of hybrid-enabling technology when producing energy to meet consumption demand, we assume energy producing firms follow the Bertrand game paradigm. In the presence of government subsidy for the development and sustainability of renewable energy, tax on pollution created by energy producing firms will motivate them to undertake Research & Development (R&D) measures to improve hybrid enabling technologies to further reduce the level of carbon pollution. As a result, from an economic point of view it is an interesting question to examine the Bertrand-Nash equilibrium under such a dynamic environment. This chapter examines this concept via a Stochastic Differential game paradigm.

Many researchers have applied game theory to study carbon reduction behavior in electricity markets. In [6], the Cournot equilibria in an oligopolistic electricity market subject to a linear demand function is examined. In [7], the power suppliers bidding behavior is evaluated under the supply function equilibrium (SFE) paradigm, where the market power of an independent system operator (ISO) is modeled as a bi-level multi-objective problem. In [8], the equilibrium strategies in random-demand procurement auctions in the electricity market is obtained and presented a method for explicit calculation of the bid strategies is presented. [9] proposed a Nash bargaining game model to examine how governments can determine the taxes and subsidies in a competitive electricity market whilst achieving their environmental objectives. In [10–14], the role of government as a leading player who intervenes in competitive electricity markets to promote environmental protection is evaluated. In [15], the role of government, when managing environmental sustainability in a complete electricity market in a Stackelberg game paradigm is examined. In [16], a more robust trans-boundary industrial pollution reduction strategy for global emission collaborations is presented. The dynamics of each country's quantity of pollution is modeled as a Brownian motion with Jumps to capture the systematic jumps caused by surprise effects arising from policy uncertainties within the economy. However, a crucial limitation within many of these environmental policy models, is that technological change is incorporated as an exogenous variable and does not consider the role of endogenous hybrid-enabling technology or other technological breakthroughs, hence limiting the dynamics of these models.

We quantify an optimal level of subsidy for the sharing of hybrid-enabling technology in an energy market under a Bertrand-duopoly game. We formulate a Stochastic Differential Game (SDG) to analyze the stability of the Stackelberg, Nash and cooperative equilibria via a feedback control strategy. We then adopt limit expectation and variance of the improvement degree to identify the influence of external environment limitations on the decision maker. We show that the game depends on its parameters and the equilibria chosen. We consider an electricity market composed of power plants I and II, with each one having the choice between fossil fuels (F) (e.g., natural gas, petroleum or coal) and renewable sources (R) (e.g., biomass, solar, wind, wave, geothermal or hydroelectric). Such hybrid power plants play a crucial ameliorating role in managing the long-standing problem of climate change and ensure immediate reliability and affordability of energy production, whilst reducing Greenhouse Gas (GHG) emissions.

In this model, we consider a Bertrand duopoly game for two power plants under endogenous hybrid-enabling technology. In the first stage the matrix of prices $\left(p_{ij}\right)_{ij\in\{F,R\}}$, (where $\left(p_{ij}\right)$ is the price of energy i, given that the opponent player's energy j) is determined as a Nash equilibrium of the game where each player wants to optimize his/her demand. We then search for the Nash equilibria, and the optimal proportions that maximizes the $(\Pi)_{ij}, ij\in\{F,R\}$ subject to the source type of energy that has been used. Once all these parameters have been fixed, the game becomes dynamic due to the evolution of a hybrid-enabling technology level $K(t)$, prompted by Research and Developments (R&D) measures undertaken by each power plant. Hence, each player must fix a time-dependent effect level associated with this hybrid-enabling technology. In doing so, this study makes the following contributions to existing game theory/energy economics literature:

i. stochastic endogenous hybrid-enabling technology innovation is introduced into a two-player stochastic differential game with random interference factors, which capture uncertain external environment factors and the internal limitations within the shared hybrid-enabling technology decision process. In doing so, we provide a framework to quantify the impacts of market power on prices.

ii. we show that both power plants invest in R&D measures and that the limit of expectation and variance of the improvement degree can be applied to identify the influence of random factors.

iii. mathematically, we show that the issue of the game depends on the parameters of the game and the type of equilibrium one considers.

iv. by applying the HJB equation we obtain the optimal effort level and the optimal level of subsidy for sharing hybrid-enabling technology via feedback equilibrium strategies whilst examining the Stackelberg equilibria, Nash equilibria and cooperative equilibria under Bertrand duopoly.

v. we reveal that for a given level of payoff distribution the Stackelberg equilibria under endogenous hybrid-technology innovation and the sharing paradigm dominate the Nash equilibria.

vi. we show that in Stackelberg and Nash games, optimal hybrid-enabling technology innovation is proportional to the government subsidy, but the variance improvement degree of the Stackelberg game is different to the results of the Nash game.

vii. our characterization of the short run price competition by strategic supplies for renewable and fossil resources, provides a more robust model than that presented by Bertrand-Edgworth, in which price competition with fixed (endogenous) capacities was used.

viii. our model shows that robust cost-reducing R&D investments with effective hybrid-enabling technology innovation strengthens an innovator's competitive position and the Stackelberg structure emerges as an equilibrium outcome, allowing each power plant to optimally use energy sources to produce electricity while maximizing their payoffs.

Therefore, under a Stochastic Differential Game (SDG) paradigm with uncertainty, each power plant can optimally use energy sources to produce electricity while maximizing their payoffs. Each power plant is capable of using fossil fuels (F) and renewable sources (R) to produce electricity at any time. To maintain the generality of the proposed model, this model is not limited to a specific energy source. Hence, the terms "F" and "R", are used throughout the paper. On the other hand the government encourages power plants to conform to a maximum accepted level of carbon emissions through strategies such as the imposition of tariffs on polluters as well as incentives for those who choose to undertake R&D measures to reduce their emission levels in order to maintain environmental sustainability. R&D spending is costly, and the presumption is that R&D spending is somehow connected to increased innovation, revenue growth and profits.

In recent years, researchers have incorporated the theory of SDGs, originated from [14, 17–20] to analyzed environmental issues. Especially [21] analyzed (two player) zero-sum stochastic differential games in a rigorous way, and proved that the upper and lower value functions of such games satisfy the dynamic programming principle whilst being the unique viscosity solutions of their associated Hamilton-Jacobi-Bellman-Isaacs equations.

In Section 2, the proposed model and elements of evolutionary game theory are presented. In Section 3 by implementing the Stackelberg game we examine feedback Stackelberg equilibria, optimal level of subsidy for the shared hybrid-enabling technology from its counterpart and the limit of expectation and variance. In Section 4 by implementing a Nash game we examine feedback Nash equilibria and the limit of expectation and variance under hybrid-enabling technology. In Section 5 by implementing a cooperative game we examine feedback equilibria and the limit of expectation and variance under hybrid-enabling technology. In Section 6, comparative analysis of equilibrium results are described. Section 7 concludes the study. Appendix at the end of the chapter contains proofs.

2. Model setup

We propose that the production process of electricity leads to emissions and is proportional to the power industry's use of energy source. We assume that there are two power plants (Player I) and (Player II) in the energy market and each power plant is capable of using *fossil fuels* (F) and *renewable sources* (R) to generate power at any given time t. To reduce the level of Green House Gas (GHG)-emissions into the atmosphere (accordance with [22] Protocol), the government will set a maximum emission quantitative level, that is directly linked to the power industry's use of energy source F, when producing electricity. Government encourages the power industry to undertake necessary hybrid-enabling technology to reduce their GHG-emission levels to the maximum accepted quantitative level, $\overline{\eta}^F$, and improve efficiency in renewables. We assume that the power plants change their strategies over time based on payoff comparisons based on hybrid-enabling technological advances. This contradicts with classical non cooperative game theory that analyzes how rational players will behave through static solution concepts such as the Nash equilibrium (NE) (i.e., a strategy choice for each player whereby no individual has a unilateral incentive to change his or her behavior).

Under the theory of evolutionary games, the production strategies in the *absence of any superior hybrid-enabling technological advances*, allows the power plants to play a symmetric two-person 2×2 bi-matrix game. Thus, for each power plant, we define the set Σ as its pure strategy given by the set of non-negative prices $[0, \infty)$. According to the Bertrand game all firms setting the lowest price will split market demand equally (Hotelling type) and the profit can be calculated subject to the electricity prices and the associated cost functions.

Then each iteration of an evolutionary game, where two matched power plants in accordance with Bertrand paradigm compete with each market and play a one-shot non-zero-sum game, represents the benchmark game of the population. If $\left(p_{ij}, p_{ji}\right)$ is the matrix of prices of power plants, respectively, then via Proposition 1 (given below), it will allow us to derive Nash equilibria of prices for these two matched power plants. On the demand side we assume that the preferences are quadratic as in [23].

We define the continuous demand function (D_{ij}), for each power plant as

$$D_{ij} = a_{ij} - \beta_{ij}\left(p_{ij} + \tau_i\right) + \gamma_{ij}\left(p_{ji} + \tau_j\right), \quad i,j \in \{F, R\} \tag{1}$$

where D_{ij} is the demand function for the power plants employing the energy source $i \in \{F, R\}$ against the power plant which use the energy source $j \in \{F, R\}$. τ_i is the tariff imposed by government subject to the power source i. For example government impose a tariff-rate quota (TRQs) (τ_F), for *fossil fuels* (F) and a feed-in-tariff (FITs) (τ_R), for *renewable sources* (R). p_{ij} is the electricity price of the power plant that uses the energy source i, versus the power plant that employs the energy source j. $a_{ij} > 0$, is the constant market base for the power plant that employs the energy source i versus the one which use the energy source j. The parameters $\beta_{ij} > 0$ and $\gamma_{ij} > 0$, are independent constants that captures the demand sensitivity of a power plant subject to its own price β_{ij} and its rival's price γ_{ij}. Eq. (1), concludes that the goods in the market are gross substitutes and that the demand function D_{ij}, is increasing in the price of the rival firm p_{ji}.

The government's tariff policy for the power plants with respect to their source of energy for long-time periods are transparent, and this information is available to the public. Therefore, it is assumed that the competitive power plants follow the government's financial legislation, having the capability and technological skills to produce electricity from specific sources at any given time to meet energy demand. Then for each time-period the power plant will consider the tariff-rate quota or feed-in-tariff and adopt a pricing strategy for the selected energy source. Hence, we conclude that the production rate of the power plants is equal to the corresponding demand rates with a negligible internal consumption and waste rate.

To apply the Backward induction technique to investigate the equilibrium prices, demand, and profits, we define the profit function for each power plant as

$$\begin{aligned}\Pi_{ij} &= \left(p_{ij} - C_i - v_i\right)D_{ij} - F_i \\ &= \left(p_{ij} - C_i - v_i\right)\left(a_{ij} - \beta_{ij}\left(p_{ij} + \tau_i\right) + \gamma_{ij}\left(p_{ji} + \tau_j\right)\right) - F_i,\end{aligned} \tag{2}$$

where $i,j \in \{F,R\}$ and $C_i > 0$ is the unit production cost of the power plant when using energy source i. $v_i > 0$, for any additional R&D unit cost for undertaking hybrid-enabling technology, for a power plant that rely on an energy source i, ($F_i > 0$ is the initial setup cost of the power plants when using the energy source i). We also assume that $\left(p_{ij} - C_i - v_i\right) > 0$. The firms' technologies are represented by their reduced cost functions. This assumes that all factor markets are perfectly competitive and – both here and in the models of imperfect competition in the output market – are not influenced by any strategic behavior of the firms in other markets. We will make alternative assumptions about those technologies. In the first assumption, pollution is proportional to output and firms do not have any further abatement technologies.

Proposition 1. *The equilibrium price for the power plants under* (τ_F, τ_R), *is given as* $p_{ij} = \Lambda_{ij} + C_i + v_i$.

Proof. Via the first order conditions of the profit function (Eq. (2)), obtain

$$\begin{aligned}\frac{\partial \Pi_{ij}}{\partial p_{ij}} &= a_{ij} - 2\beta_{ij}\left(p_{ij} - C_i - v_i\right) + \gamma_{ij}\left(p_{ji} - C_j - v_j\right) + \gamma_{ij}(\tau_j + C_j + v_j) \\ &\quad -\beta_{ij}(\tau_i + C_i + v_i) = 0.\end{aligned} \tag{3}$$

Defining $\Lambda_{ij} = p_{ij} - C_i - v_i$ and using Λ_{ij} and Λ_{ji}, rewrite the first order conditions as:

$$\begin{cases} 2\beta_{ij}\Lambda_{ij} - \gamma_{ij}\Lambda_{ji} &= a_{ij} + \gamma_{ij}(\tau_j + C_j + v_j) - \beta_{ij}(\tau_i + C_i + v_i), \\ 2\beta_{ji}\Lambda_{ji} - \gamma_{ji}\Lambda_{ij} &= a_{ji} + \gamma_{ji}(\tau_i + C_i + v_i) - \beta_{ji}(\tau_j + C_j + v_j). \end{cases} \tag{4}$$

Simultaneously solving Eq. (4), obtain

$$\Lambda_{ij} = \frac{2\beta_{ji}a_{ij} + \gamma_{ij}a_{ji} + \beta_{ji}\gamma_{ij}(\tau_j + (C_j + v_j)) + \left(\gamma_{ij}\gamma_{ji} - 2\beta_{ji}\beta_{ij}\right)(\tau_i + (C_i + v_i))}{\left(4\beta_{ji}\beta_{ij} - \gamma_{ij}\gamma_{ji}\right)} \tag{5}$$

such that $\beta_{ji}\beta_{ij} \neq \frac{\gamma_{ij}\gamma_{ji}}{4}$. p_{ij}^* and p_{ji}^* are obtained via $p_{ij} = \Lambda_{ij} + C_i + v_i$ and are the optimum prices if the profit functions are concave on p_{ij} and on p_{ji}. Then via the second order conditions, obtain the maximum point in the set as: $\frac{\partial^2 \Pi_{ij}}{\partial p_{ij}^2} = -2\beta_{ij} < 0$. Since $\beta_{ij} > 0$, implies that the second derivative of the profit function in equilibrium is negative confirming that the profit function is concave at this point.

Proposition 2. *At equilibrium prices the power plant's demand and profit under* (τ_F, τ_R), *can be obtained as*

$$\begin{aligned} D_{ij}^* &= \beta_{ij}\Lambda_{ij}^* = \beta_{ij}\left(\theta_{ij} + \omega_{ij}\tau_i^* + \chi_{ij}\tau_j^*\right), \\ \Pi_{ij}^* &= \beta_{ij}\left(\Lambda_{ij}^*\right)^2 - F_i = \beta_{ij}\left(\theta_{ij} + \omega_{ij}\tau_i^* + \chi_{ij}\tau_j^*\right)^2 - F_i, \end{aligned} \tag{6}$$

where

$$\theta_{ij} = \frac{\left(2\beta_{ji}a_{ij} + \gamma_{ij}a_{ji} + \beta_{ji}\gamma_{ij}\left(C_j + v_j\right) + \left(\gamma_{ij}\gamma_{ji} - 2\beta_{ji}\beta_{ij}\right)\left(C_i + v_i\right)\right)}{\left(4\beta_{ji}\beta_{ij} - \gamma_{ij}\gamma_{ji}\right)}, \tag{7}$$

$$\omega_{ij} = \frac{\left(\gamma_{ij}\gamma_{ji} - 2\beta_{ji}\beta_{ij}\right)}{\left(4\beta_{ji}\beta_{ij} - \gamma_{ij}\gamma_{ji}\right)} \text{ and } \chi_{ij} = \frac{\beta_{ji}\gamma_{ij}}{\left(4\beta_{ji}\beta_{ij} - \gamma_{ij}\gamma_{ji}\right)} \tag{8}$$

Proof. Obtain the results by substituting Λ_{ij}^* from Proposition 1 into Eqs. (1) and (2) and simplifying.

Remark 1. The only outcome where neither power plant has an incentive to deviate is when $p_{ij} = p_{ji} = c_i$, which will be the Nash or Bertrand equilibrium for the game. The intuition behind this result is that power plants will keep *undercutting* the price of its rival until price equals marginal cost. In the long run price changes with marginal cost and industry production increases with demand and falls with marginal cost. One way for a power plant to avoid the Bertrand paradox and earn economic profit in a Bertrand setting is to have a competitive cost advantage over its rival.

2.1 Production decisions of power plants with homogenous hybrid/enabling technology

Restricting ourselves to a two matched symmetric two-person bi- matrix game in random contest in a one-population evolutionary game, we define the payoff (utility) in **Table 1**.

Note 1. Power and the payoff are measured on a utility scale consistent with the power plant's preference ranking. Furthermore, [24–26] have applied symmetric two-person bi-matrix game in random contest to study evolutionary stable games.

		Power Plant II	
	Production Strategy	Fossil Fuel	Renewable Sources
Power Plant I	Fossil Fuel	(Π_{FF}, Π_{FF})	(Π_{FR}, Π_{RF})
	Renewable Sources	(Π_{RF}, Π_{FR})	(Π_{RR}, Π_{RR})

Table 1.
Bi matrix for two power plants by different energy sources.

Then via equation $\Pi_{ij}^* = \beta_{ij}\left(\Lambda_{ij}^*\right)^2 - F_i = \beta_{ij}\left(\theta_{ij} + \omega_{ij}\tau_i + \chi_{ij}\tau_j\right)^2 - F_i$, in Proposition 1, and the payoff matrix of the power plant I is given by:

$$A = \begin{bmatrix} a_{11} & a_{12} \\ a_{21} & a_{22} \end{bmatrix} = \begin{bmatrix} \Pi_{F,F} & \Pi_{F,R} \\ \Pi_{R,F} & \Pi_{R,R} \end{bmatrix} = \begin{bmatrix} \beta_{F,F}\Lambda_{F,F}^2 - F_F & \beta_{F,R}\Lambda_{F,R}^2 - F_F \\ \beta_{R,F}\Lambda_{R,F}^2 - F_R & \beta_{R,R}\Lambda_{R,R}^2 - F_R \end{bmatrix}. \tag{9}$$

Obviously the bimatrix of the power plant II, is given by:

$$A = \begin{bmatrix} a_{11} & a_{21} \\ a_{12} & a_{22} \end{bmatrix} = \begin{bmatrix} \Pi_{F,F} & \Pi_{R,F} \\ \Pi_{F,R} & \Pi_{R,R} \end{bmatrix} = \begin{bmatrix} \beta_{F,F}\Lambda_{F,F}^2 - F_F & \beta_{R,F}\Lambda_{R,F}^2 - F_R \\ \beta_{F,R}\Lambda_{F,R}^2 - F_F & \beta_{R,R}\Lambda_{R,R}^2 - F_R \end{bmatrix}. \tag{10}$$

Proposition 3. *The Nash equilibrium for the Bi-matrix game G, is given as*

$$\left(\frac{(\Pi_{R,R} - \Pi_{F,R})}{(\Pi_{F,F} - \Pi_{R,F} - \Pi_{FR} + \Pi_{R,R})}, \frac{(\Pi_{R,R} - \Pi_{F,R})}{(\Pi_{F,F} - \Pi_{F,R} - \Pi_{R,F} + \Pi_{R,R})}\right). \tag{11}$$

Proof. Suppose players I and II use mixed strategies (x,1-x) and (y,1-y), respectively, where

i. The probability that player I choosing row 1 is x and the probability that player I choosing row 2 is 1-x.

ii. The probability that player II choosing row 1 is y and the probability that player II choosing row 2 is 1-y.

Then the value of the game for Player I is

$$v_1(x,y) = xy(\Pi_{F,F}) + x(1-y)(\Pi_{F,R}) + (1-x)y(\Pi_{R,F}) + (1-x)(1-y)(\Pi_{R,R})$$

$$= ((\Pi_{F,F} - \Pi_{F,R} - \Pi_{R,F} + \Pi_{R,R})y + (\Pi_{F,R} - \Pi_{R,R}))x + ((\Pi_{R,F} - \Pi_{R,R})y + \Pi_{R,R}), \tag{12}$$

and the value of the game for Player II is

$$v_2(x,y) = xy(\Pi_{F,F}) + x(1-y)(\Pi_{R,F}) + (1-x)y(\Pi_{F,R}) + (1-x)(1-y)(\Pi_{R,R})$$

$$= ((\Pi_{F,F} - \Pi_{R,F} - \Pi_{F,R} + \Pi_{R,R})x + (\Pi_{F,R} - \Pi_{R,R}))y + ((\Pi_{R,FR} - \Pi_{R,R})x + \Pi_{R,R}). \tag{13}$$

Suppose (X, Y) yields a Nash equilibrium. Then for the given payoffs having $0 < x < 1$ implies that

$$v_1 = (\Pi_{F,F} - \Pi_{F,R} - \Pi_{R,F} + \Pi_{R,R})y + (\Pi_{F,R} - \Pi_{R,R}) = 0. \tag{14}$$

Otherwise Player I can change x slightly and do better.
Similarly, for $0 < y < 1$,

$$v_2 = (\Pi_{F,F} - \Pi_{R,F} - \Pi_{FR} + \Pi_{R,R})x + (\Pi_{F,R} - \Pi_{R,R}) = 0. \tag{15}$$

Otherwise Player II can change y slightly and do better. It follows that the unique Nash equilibrium (x,y), has

$$\left(\frac{(\Pi_{R,R} - \Pi_{F,R})}{(\Pi_{F,F} - \Pi_{R,F} - \Pi_{FR} + \Pi_{R,R})}, \frac{(\Pi_{R,R} - \Pi_{R,F})}{(\Pi_{F,F} - \Pi_{F,R} - \Pi_{R,F} + \Pi_{R,R})}\right). \tag{16}$$

Remark 1. Since the power plants plays a symmetric two person bimatrix game G, having two pure strategies $\Pi_{F,F} \neq \Pi_{R,F}, \quad \Pi_{R,R} \neq \Pi_{FR}$, imply that G, has an evolutionary stable strategy. Then the Nash equilibrium is an outcome in which the strategy chosen by each player is the best reply to the strategy chosen by the other. This best reply strategy yields the highest payoff to the player choosing it, given the strategy chosen by the co-player, [27, 28].

2.2 Production decisions of power plants under endogenous hybrid/enabling technological advances

Both players will undertake R&D measures on hybrid-enabling technology to ensure immediate reliability and affordability in energy production whilst reducing GHG-emissions. We assume that the strategic effects implemented by power plant I (Player I), has improved hybrid-enabling technology to generate energy and utilize energy sources in a much efficient way. This gives a superior advantage to power plant I overpower plant II (Player II) and both power plants are rational to maximize their profits. Although Power plant II has heterogeneous resources to hybrid-enabling technology, from a practical point of view it is logical for power plant I to share this technology with power plant II, because the price competition is typically characterized by a second-mover advantage. Many researchers have investigated the effects of these commitments in Cournot, Bertrand and Stackelberg setups. See [29–31]. Due to the government incentives, tariff-rate quota, feed-in-tariff and R&D incentive measures, the power companies will be competitive to improve their efficiency. Let $L^R(t)$ denotes the *R&D effort level* of technological improvements on renewable sources at time t, and $L^F(t)$ denotes the *R&D effort level* of technological improvements on fossil fuel at time t, of Player I. $\tilde{L}^R(t)$ denotes the *R&D effort level* of technological improvements on renewable sources at time t, and $\tilde{L}^F(t)$ denotes the *R&D effort level* of technological improvements on fossil fuel at time t, of Player II. For, further consideration, the sharing cost of advanced hybrid-enabling technology (Player I) and inferior hybrid-enabling technology (Player II) is denoted as $C_{\mathrm{I}}(t)$ and $C_{\mathrm{II}}(t)$, which are the quadratic functions of the effect level of Player I and Player II at time t, respectively. Consider

$$C_{\mathrm{I}}\left(L^R(t), L^F(t), t\right) = \frac{1}{2}\left(\beta^R(t)\left(L^R(t)\right)^2 + \beta^F(t)\left(L^F(t)\right)^2\right), \tag{17}$$

and

$$C_{\mathrm{II}}\left(\tilde{L}^R(t), \tilde{L}^F(t), t\right) = \frac{1}{2}\left(\tilde{\beta}^R(t)\left(\tilde{L}^R(t)\right)^2 + \tilde{\beta}^F(t)\left(\tilde{L}^F(t)\right)^2\right), \tag{18}$$

where $0 < \left(\beta^R(t), \beta^F(t), \tilde{\beta}^R(t), \tilde{\beta}^F(t)\right) \leq 1$ and lower the $\left(\beta^R(t), \beta^F(t), \tilde{\beta}^R(t), \tilde{\beta}^F(t)\right)$, more effective is the technological development.

Let $K(t)$ denote the evolution of the hybrid-enabling technology at time t, due to R&D collaborative innovation system of Player I and Player II at time t. The dynamics of hybrid-technology is governed by the stochastic differential equation (SDE):

$$\begin{cases} dK(t) = \left[\vartheta_1(t)\left(L^R(t), L^F(t)\right) + \vartheta_2(t)\left(\tilde{L}^R(t), \tilde{L}^F(t)\right) - \xi K(t)\right]dt + \varphi\sqrt{K}dW(t) \\ K(0) = K_0 > 0. \end{cases} \tag{19}$$

$\xi \in (0, 1]$, is the attenuation coefficient of hybrid-enabling technology. Let $\vartheta_1(t) = \left(\vartheta_1^R(t) + \vartheta_1^F(t)\right)$ and $\vartheta_2(t) = \left(\vartheta_2^R(t) + \vartheta_2^F(t)\right)$ denote the influence of the effort level of hybrid-enabling technology sharing on collaboration innovation between Player I and Player II, at time t. $W(t)$ is a standard Brownian motion and $\varphi\left(\sqrt{K}(t)\right)$ random interference factor on hybrid-enabling technology.

Let $\Pi(t)$ denotes the total profit under the hybrid-enabling technology system at time t. Let $(\alpha_1(t), \alpha_2(t))$ and $(\beta_1(t), \beta_2(t))$ denote the influence of the effort level hybrid-enabling technology on the total profit of Player I and player II, respectively, at time t, namely, the marginal return coefficient of hybrid-enabling technology. Total profit function can be expressed as:

$$\Pi(t) = \left(\alpha_1(t)L^R(t) + \alpha_2(t)L^F(t)\right) + \left(\beta_1(t)\tilde{L}^R(t) + \beta_2(t)\tilde{L}^F(t)\right) + (\Gamma + \delta)K(t), \quad (20)$$

where

$$\alpha_1(t) = \frac{\Pi_{R,R}(t)}{\Pi_{F,F}(t) - \Pi_{R,F}(t) - \Pi_{F,R}(t) + \Pi_{R,R}(t)}, \quad (21)$$

$$\alpha_2(t) = \frac{-\Pi_{F,R}(t)}{\Pi_{F,F}(t) - \Pi_{R,F}(t) - \Pi_{F,R}(t) + \Pi_{R,R}(t)}, \quad (22)$$

$$\Gamma = \Gamma_{(\mathrm{I})} + \Gamma_{(\mathrm{II})}, \delta = \delta_{(\mathrm{I})} + \delta_{(\mathrm{II})}, \text{and}$$

$$\beta_1(t) = \frac{\Pi_{R,R}(t)}{\Pi_{F,F}(t) - \Pi_{R,F}(t) - \Pi_{F,R}(t) + \Pi_{R,R}(t)}, \quad (23)$$

$$\beta_2(t) = \frac{-\Pi_{R,F}(t)}{\Pi_{F,F}(t) - \Pi_{R,F}(t) - \Pi_{F,R}(t) + \Pi_{R,R}(t)}. \quad (24)$$

Γ is the influence of the hybrid-enabling technology innovation on total revenue $\delta \in (0, 1]$; δ is the total government subsidy coefficient of hybrid-enabling technology based on increments of advances in hybrid-enabling technology.

Proposition 4. *At least one of the Power Plants has a second mover advantage.*

Proof. Demand function $D_{ij}\left(p_{ij}, p_{ji}\right) > 0$, given by Eq. (4), is twice continuously differentiable and

$$\frac{\partial D_{ij}\left(p_{ij}, p_{ji}\right)}{\partial p_{ij}} = -\beta_{ij} < 0, and \frac{\partial D_{ij}\left(p_{ij}, p_{ji}\right)}{\partial p_{ji}} = \gamma_{ij} > 0 \forall \left(p_{ij}, p_{ji}\right) \in P_{\mathrm{I}} \times P_{\mathrm{II}}. \quad (25)$$

The first inequality says that each demand is downward sloping in own price, and the second that goods are substitutes (each demand increases with the price of the other good). [32] shows that in case of symmetric firms, there is a second-mover (first-mover) advantage for both players when each profit function is strictly concave in own action and strictly increasing (decreasing) in rival's action, and reaction curves are upward (downward) sloping.

Then a sufficient condition on the super-modularity of the profit function is obtained via the profit function Π_{ij}, given by Eq. (4):

$$\left[\frac{\partial D_{ij}\left(p_{ij}, p_{ji}\right)}{\partial p_{ji}} + \left(p_{ij} - C_i - v_i\right)\frac{\partial^2 D_{ij}\left(p_{ij}, p_{ji}\right)}{\partial p_{ji} p_{ij}}\right] E(K(t)) > 0, \quad (26)$$

where E is the expectations. The main implication of this is that it leads to reaction correspondences that are non-decreasing (in the sense that each selection is non-decreasing) but need not be single-valued or continuous. This has a very appealing and precise interpretation: The price elasticity of Power Plant i's demand increases in the rival's price, [33]. This is a very intuitive and general condition, though clearly not a universal one. It is satisfied in particular if $\frac{\partial^2 D_{ij}(p_{ij}, p_{ji})}{\partial p_{ji} p_{ij}} > 0$, if a higher price by a Power Plant's rival does not lower the responsiveness of the Power plant's demand to a change in own price.

We further assume that the total revenue is allocated between two players and $\theta(t)$ is the payoff distribution coefficient of player I at time t and $\theta(t) \in [0, 1]$. Although Player II has heterogeneous resources of hybrid-enabling technology, Player I can produce electricity more efficiently with lower GHG-emission, ensure immediate reliability and affordability in energy production. Then Player II, can acquire practical outcomes of this hybrid-enabling technological advances. To promote the hybrid-enabling technology, Player II (leader) determine an optimal sharing effort level and an optimal subsidy. Then Player I (follower) choose their optimal sharing effort level according to the optimal sharing effort level and subsidy. This leads to a Stackelberg equilibrium. Let $\omega(t) = (\omega_1(t), \omega_2(t))$, denote the subsidy for hybrid-enabling technology, with Player II willing to pay to Payer I under collaboration. The objective functions of power plant I and power plant II satisfy the following partial differential equations

$$
\begin{aligned}
&J_{(\mathrm{I})}(K_0)\\
&= \max_{\{L_S^R, L_S^F\} \geq 0} E\left\{\int_0^\infty e^{-\rho_1 t}\left[\theta(t)\left(\alpha_1(t)L^R(t) + \alpha_2(t)L^F(t) + \beta_1(t)\tilde{L}^R(t) + \beta_2(t)\tilde{L}^F(t)\right.\right.\right.\\
&\left.\left.\left.+(\Gamma+\delta)K(t)\right)\right) - \frac{1}{2}\beta^R(t)(1-\omega_1)\left(L^R(t)\right)^2 - \frac{1}{2}\beta^F(t)(1-\omega_2)\left(L^F(t)\right)^2\right]dt\right\},
\end{aligned} \tag{27}
$$

and

$$
\begin{aligned}
J_{(\mathrm{II})}(K_0) = &\max_{\{\tilde{L}_S^R, \tilde{L}_S^F, \omega(t)\} \geq 0} E\left\{\int_0^\infty e^{-\rho_2 t}\left[(1-\theta(t))\left(\alpha_1(t)L^R(t) + \alpha_2(t)L^F(t) + \beta_1(t)\tilde{L}^R(t)\right.\right.\right.\\
&\left.+\beta_2(t)\tilde{L}^F(t) + (\Gamma+\delta)K(t)\right) - \frac{1}{2}\tilde{\beta}^R(t)\left(L^R(t)\right)^2 - \frac{1}{2}\tilde{\beta}^F(t)\left(\tilde{L}^F(t)\right)^2\\
&\left.\left.-\frac{1}{2}\omega_1\beta^R(t)\left(L^R(t)\right)^2 - \frac{1}{2}\omega_2\beta^F(t)\left(L^F(t)\right)^2\right]dt\right\},
\end{aligned} \tag{28}
$$

where ρ_1 and ρ_2 are the discount rates of Player I and Player II, respectively. In this feedback control strategy $L_S^R(t) \geq 0$, $L_S^F(t) \geq 0$, L S R (t) ≥ 0, $\tilde{L}_S^R(t) \geq 0$ and $\tilde{L}_S^F(t) \geq 0$, are the control variables and $\omega(t) = (\omega_1(t), \omega_2(t)) \in (0, 1)$. $K(t) > 0$ is the state variable. In feedback control process, it is assumed that players at every point in time have access to the current system and can make decisions accordingly to that state. Consequently, the players can respond to any disturbance in an optimal way. Hence, feedback strategies are robust for deviations and players can react to disturbances during the evolution of the game and adapt their actions accordingly, [34].

3. A Stackelberg game under heterogeneous technology

Theory of strong Stackelberg reasoning is an improved version of an earlier theory [35], which provides an explanation of coordination in all dyadic (two-player) common interest games. It provides an explanation of why players tend to choose strategies associated with a payoff-dominant Nash equilibrium. Its distinctive assumption is that players behave as though their co-players will anticipate any strategy choice and invariably choose a best reply to it. Stackelberg strategies resulting from this form of reasoning do not form Nash equilibria. The theory makes no predictions, because a non-equilibrium outcome is inherently unstable, leaving at least one player with a reason to choose differently and thereby achieve a better payoff. Strong Stackelberg reasoning is a simple theory, according to which players in dyadic games choose strategies that would maximize their own payoffs if their co-players could invariably anticipate their strategy choices and play counter-strategies that yield the maximum payoffs for themselves. The key assumption is relatively innocuous, first because game theory imposes no constraints on players' beliefs, apart from consistency requirements, and second because the theory does not assume that players necessarily believe that their strategies will be anticipated, merely that they behave as though that is the case, as a heuristic aid to choosing the best strategy. Strong Stackelberg reasoning is, in fact, merely a generalization of the minorant and majorant models introduced by [36] and used to rationalize their solution of strictly competitive games.

To promote the sharing of hybrid-enabling technology, the Player II (the leader) determine an optimal sharing effort sharing level and an optimal subsidy scheme. Then the Player I (the follower) choose his/her optimal sharing level according to the optimal sharing effort level and subsidy. This leads to a Stackelberg equilibrium.

Proposition 5. *If above conditions are satisfied, the feedback Stackelberg leader (Player II)-follower (Player I) and equilibria is given as:*

$$L_S^R = \frac{\alpha_1(2-\theta)(\rho_2+\xi)(\rho_1+\xi) + \vartheta_1^R(\Gamma+\delta)((2-2\theta)(\rho_1+\xi)+\theta(\rho_2+\xi))}{2\beta^R(\rho_2+\xi)(\rho_1+\xi)}, \tag{29}$$

$$L_S^F = \frac{\alpha_2(2-\theta)(\rho_2+\xi)(\rho_1+\xi) + \vartheta_1^F(\Gamma+\delta)((2-2\theta)(\rho_1+\xi)+\theta(\rho_2+\xi))}{2\beta^F(\rho_2+\xi)(\rho_1+\xi)}, \tag{30}$$

$$\tilde{L}_S^R = \frac{(1-\theta)(\beta_1(\rho_2+\xi)+(\Gamma+\delta))\vartheta_2^R}{\tilde{\beta}^R(\rho_2+\xi)}, \tag{31}$$

$$\tilde{L}_S^F = \frac{(1-\theta)(\beta_2(\rho_2+\xi)+(\Gamma+\delta))\vartheta_2^F}{\tilde{\beta}^F(\rho_2+\xi)}. \tag{32}$$

where L_S^R, L_S^F are the optimal effort level of hybrid-enabling technological improvements shared on renewable sources and fossil fuel at time t by Player I, respectively. $\tilde{L}_S^R, \tilde{L}_S^F$ are the optimal effort level of technological improvements shared on renewable sources and fossil fuel at time t by Player II, respectively.

The optimal level of subsidy for sharing hybrid-enabling on renewable sources is given by

$$\omega_1 = \begin{cases} \dfrac{\alpha_1(2-3\theta)+\vartheta_1^R[2a_2-a_1]}{\alpha_1(2-\theta)+\vartheta_1^R[2a_2+a_1]}, & 0 \le \theta \le \dfrac{2}{3} \\ 0. & \text{otherwise} \end{cases} \tag{33}$$

Similarly, the optimal level of subsidy for sharing hybrid-enabling technology on fossil fuel is given by:

$$\omega_2 = \begin{cases} \dfrac{\alpha_2(2-3\theta)+\vartheta_1^F[2a_2-a_1]}{\alpha_2(2-\theta)+\vartheta_1^F[2a_2+a_1]}, & 0\le\theta\le\dfrac{2}{3} \\ 0. & \text{otherwise} \end{cases} \tag{34}$$

The optimal sharing payoff functions under hybrid-enabling technology on renewable sources and on fossil fuel for Player I and Player II are given below

$$V_S^{(\mathrm{I})}(K) = \frac{\theta(\Gamma+\delta_))}{(\rho_1+\xi)}K + b_1, \quad V_S^{(\mathrm{II})}(K) = \frac{(1-\theta)(\Gamma+\delta)}{(\rho_2+\xi)}K + b_2, \tag{35}$$

where a_1, a_2, b_1 and b_2 are given in the proof.

Proof. We define the optimal revenue functions for Player I and Player II under hybrid-enabling technology as $V_S^{(\mathrm{I})}(K)$ and $V_S^{(\mathrm{II})}(K)$, respectively, which are continuously differentiable. Applying HJB equation to $V_S^{(\mathrm{I})}(K)$, for Player I, we obtain

$$\rho_1 V_S^{(\mathrm{I})}(K) = \max_{\{L_S^R, L_S^F\}\ge 0}\Big\{\Big[\theta\Big(\alpha_1 L_S^R(t)+\alpha_2 L_S^F+\beta_1\tilde{L}_S^R+\beta_2\tilde{L}_S^F+(\Gamma+\delta)K\Big)\Big]$$

$$-\frac{1}{2}\beta^R(1-\omega_1)\left(L_S^R\right)^2-\frac{1}{2}\beta^F(1-\omega_2)\left(L_S^F\right)^2$$

$$+\frac{\partial V_S^{(\mathrm{I})}(K)}{\partial K}\Big[\vartheta_1\left(L_S^R,L_S^F\right)+\vartheta_2\left(\tilde{L}_S^R,\tilde{L}_S^F\right)-\xi K\Big]+\frac{1}{2}\frac{\partial^2 V_S^{(\mathrm{I})}(K)}{\partial K^2}\varphi^2(K)\Big\}. \tag{36}$$

Via the first order conditions, we obtain the optimal values $\left(L_S^R, L_S^F\right)$ for Player I as:

$$L_S^R = \frac{\theta\alpha_1 + V'^{(\mathrm{I})}_S(K)\vartheta_1^R}{\beta^R(1-\omega_1)}, \tag{37}$$

$$L_S^F = \frac{\theta\alpha_2 + V'^{(\mathrm{I})}_S(K)\vartheta_1^F}{\beta^F(1-\omega_2)}, \tag{38}$$

where $\frac{\partial V_S^{(\mathrm{I})}(K)}{\partial K} \equiv V'^{(\mathrm{I})}_S(K)$. The optimal sharing revenue function, $V'^{(\mathrm{II})}_S(K)$, for Player II and the associated HJB equation is

$$\rho_2 V_S^{(\mathrm{II})}(K) = \max_{\{\tilde{L}_S^R, \tilde{L}_S^F\}\ge 0}\Big\{\Big[(1-\theta)\Big(\alpha_1 L_S^R+\alpha_2 L_S^F+\beta_1\tilde{L}_S^R+\beta_2\tilde{L}_S^F+(\Gamma+\delta)K\Big)\Big]$$

$$-\frac{1}{2}\tilde{\beta}^R\left(L_S^R(t)\right)^2-\frac{1}{2}\tilde{\beta}^F\left(\tilde{L}_S^F\right)^2-\frac{1}{2}\omega_1\beta^R\left(L_S^R\right)^2-\frac{1}{2}\omega_2\beta^F\left(L_S^F\right)^2$$

$$+\frac{\partial V_S^{(\mathrm{II})}(K)}{\partial K}\Big[\vartheta_1\left(L_S^R,L_S^F\right)+\vartheta_2\left(\tilde{L}_S^R,\tilde{L}_S^F\right)-\xi K\Big]+\frac{1}{2}\frac{\partial^2 V_S^{(\mathrm{II})}(K)}{\partial K^2}\varphi^2(K)\Big\}. \tag{39}$$

Substituting the results of Eqs. (37) and (38) into Eq. (39), obtain

$$\rho_2 V_S^{(II)}(K) = \max_{\{\tilde{L}_S^R, \tilde{L}_S^F\} \geq 0} \left\{ \left[(1-\theta) \left(\frac{\alpha_1 \left(\theta\alpha_1 + V'^{(I)}_S(K)\vartheta_1^R \right)}{\beta^R(1-\omega_1)} + \frac{\alpha_2 \left(\theta\alpha_2 + V'^{(I)}_S(K)\vartheta_1^F \right)}{\beta^F(1-\omega_2)} \right. \right. \right.$$
$$\left. + \beta_1 \tilde{L}_S^R + \beta_2 \tilde{L}_S^F + (\Gamma + \delta)K \right) - \frac{1}{2}\tilde{\beta}^R \left(\tilde{L}_S^R(t) \right)^2 - \frac{1}{2}\tilde{\beta}^F \left(\tilde{L}_S^F \right)^2$$
$$\left. - \frac{1}{2}\omega_1 \beta^R \left(\frac{\theta\alpha_1 + V'^{(I)}_S(K)\vartheta_1^R}{\beta^R(1-\omega_1)} \right)^2 - \frac{1}{2}\omega_2 \beta^F \left(\frac{\theta\alpha_2 + V'^{(I)}_S(K)\vartheta_1^F}{\beta^F(1-\omega_2)} \right)^2 \right]$$
$$+ \frac{\partial V_S^{(II)}(K)}{\partial K} \left[\frac{\vartheta_1^R \left[\theta\alpha_1 + V'^{(I)}_S(K)\vartheta_1^R \right]}{\beta^R(1-\omega_1)} + \frac{\vartheta_1^F \left[\theta\alpha_2 + V'^{(I)}_S(K)\vartheta_1^F \right]}{\beta^F(1-\omega_2)} + \vartheta_2 \left(\tilde{L}_S^R, \tilde{L}_S^F \right) - \xi K \right]$$
$$\left. + \frac{1}{2} \frac{\partial^2 V_S^{(II)}(K)}{\partial K^2} \varphi^2(K) \right\}. \tag{40}$$

Via the first order conditions of (Eq. (40)), we obtain the optimal values $\left(\tilde{L}^R, \tilde{L}^F \right)$ for Player II as:

$$\tilde{L}_S^R = \frac{(1-\theta)\beta_1 + V'^{(II)}_S(K)\vartheta_2^R}{\tilde{\beta}^R}, \tag{41}$$

$$\tilde{L}_S^F = \frac{(1-\theta)\beta_2 + V'^{(II)}_S(K)\vartheta_2^F}{\tilde{\beta}^F}. \tag{42}$$

And the optimal value for (ω_1, ω_2)

$$\omega_1 = \frac{\alpha_1(2-3\theta) + \vartheta_1^R \left[2V'_S{}^{(II)}(K) - V_S{}'^{(I)}(K) \right]}{\alpha_1(2-\theta) + \vartheta_1^R \left[2V'_S{}^{(II)}(K) + V_S{}'^{(I)}(K) \right]}, \tag{43}$$

and

$$\omega_2 = \frac{\alpha_2(2-3\theta) + \vartheta_1^F \left[2V'_S{}^{(II)}(K) - V_S{}'^{(I)}(K) \right]}{\alpha_2(2-\theta) + \vartheta_1^F \left[2V'_S{}^{(II)}(K) + V_S{}'^{(I)}(K) \right]}. \tag{44}$$

Hence, the solution of the HJB equation is an unary function with K (K as the independent variable), we define $V_S^{(I)} = a_1 K + b_1$ and $V_S^{(II)} = a_2 K + b_2$, where a_1, b_1, a_2, and b_2 are constants that need to be solved. Simplifying (Eq. (39)), obtain:

$$\rho_1 V_S^{(I)}(K) = \theta \left(\alpha_1 \left(\frac{\theta\alpha_1 + a_1\vartheta_1^R}{\beta^R(1-\omega_1)} \right) + \alpha_2 \left(\frac{\theta\alpha_2 + a_1\vartheta_1^F}{\beta^F(1-\omega_2)} \right) + \beta_1 \left(\frac{(1-\theta)\beta_1 + a_2\vartheta_2^R}{\tilde{\beta}^R} \right) \right. \tag{45}$$

$$
\begin{aligned}
&+\beta_2\left(\frac{(1-\theta)\beta_2+a_2\vartheta_2^F}{\tilde{\beta}^F}\right)+(\Gamma+\delta)K\Bigg)-\frac{1}{2}\beta^R(1-\omega_1)\left(\frac{\theta\alpha_1+a_1\vartheta_1^R}{\beta^R(1-\omega_1)}\right)^2\\
&-\frac{1}{2}\beta^F(1-\omega_2)\left(\frac{\theta\alpha_2+a_1\vartheta_1^F}{\beta^F(1-\omega_2)}\right)^2-\xi K a_1\\
&+\left[\frac{\vartheta_1^R\left[\theta\alpha_1+a_1\vartheta_1^R\right]}{\beta^R(1-\omega_1)}+\frac{\vartheta_1^F\left[\theta\alpha_2+a_1\vartheta_1^F\right]}{\beta^F(1-\omega_2)}\right]a_1\\
&+\left[\frac{\vartheta_2^R\left[(1-\theta)\beta_1+a_2\vartheta_2^R\right]}{\tilde{\beta}^R}+\frac{\vartheta_2^F\left[(1-\theta)\beta_2+a_2\vartheta_2^F\right]}{\tilde{\beta}^F}\right]a_1,
\end{aligned}
\tag{46}
$$

and simplifying (Eq. (40)), obtain:

$$
\begin{aligned}
\rho_2 V_S^{(\mathrm{II})}(K)=(1-\theta)&\left(\frac{\alpha_1(\theta\alpha_1+a_1\vartheta_1^R)}{\beta^R(1-\omega_1)}+\frac{\alpha_2(\theta\alpha_2+a_1\vartheta_1^F)}{\beta^F(1-\omega_2)}+\beta_1\left(\frac{(1-\theta)\beta_1+a_2\vartheta_2^R}{\tilde{\beta}^R}\right)\right.\\
&\left.+\beta_2\left(\frac{(1-\theta)\beta_2+a_2\vartheta_2^F}{\tilde{\beta}^F}\right)+(\Gamma+\delta)K\right)-\frac{1}{2}\tilde{\beta}^R\left(\frac{(1-\theta)\beta_1+a_2\vartheta_2^R}{\tilde{\beta}^R}\right)^2\\
&-\frac{1}{2}\tilde{\beta}^F\left(\frac{(1-\theta)\beta_2+a_2\vartheta_2^F}{\tilde{\beta}^F}\right)^2-\frac{1}{2}\omega_1\beta^R\left(\frac{\theta\alpha_1+a_1\vartheta_1^R}{\beta^R(1-\omega_1)}\right)^2-\frac{1}{2}\omega_2\beta^F\left(\frac{\theta\alpha_2+a_1\vartheta_1^F}{\beta^F(1-\omega_2)}\right)^2\\
&+\left[\frac{\vartheta_1^R\left[\theta\alpha_1+a_1\vartheta_1^R\right]}{\beta^R(1-\omega_1)}+\frac{\vartheta_1^F\left[\theta\alpha_2+a_1\vartheta_1^F\right]}{\beta^F(1-\omega_2)}\right]a_2-\xi K a_2\\
&+\left[\frac{\vartheta_2^R\left[(1-\theta)\beta_1+a_2\vartheta_2^R\right]}{\tilde{\beta}^R}+\frac{\vartheta_2^F\left[(1-\theta)\beta_2+a_2\vartheta_2^F\right]}{\tilde{\beta}^F}\right]a_2.
\end{aligned}
\tag{47}
$$

This implies that,

$$
a_1=\frac{\theta(\Gamma+\delta)}{(\rho_1+\xi)},\quad b_1=\frac{\Phi_1}{\rho_1},\quad a_2=\frac{(1-\theta)(\Gamma+\delta)}{(\rho_2+\xi)},\quad b_2=\frac{\Phi_2}{\rho_2},\tag{48}
$$

where

$$
\begin{aligned}
\Phi_1=&\left(\alpha_1\theta-\frac{(\theta\alpha_1+a_1\vartheta_1^R)}{2}+\vartheta_1^R a_1\right)\left(\frac{\theta\alpha_1+a_1\vartheta_1^R}{\beta^R(1-\omega_1)}\right)\\
&+\left(\alpha_2\theta-\frac{(\theta\alpha_2+a_1\vartheta_1^F)}{2}+\vartheta_1^F a_1\right)\left(\frac{\theta\alpha_2+a_1\vartheta_1^F}{\beta^F(1-\omega_2)}\right)\\
&+\left(\beta_1\theta+\vartheta_2^R a_1\right)\left(\frac{(1-\theta)\beta_1+a_2\vartheta_2^R}{\tilde{\beta}^R}\right)\\
&+\left(\beta_2\theta+\vartheta_2^F a_2\right)\left(\frac{(1-\theta)\beta_2+a_2\vartheta_2^F}{\tilde{\beta}^F}\right)>0,
\end{aligned}
\tag{49}
$$

and

$$\Phi_2 = \left((1-\theta)\alpha_1 - \frac{\omega_1(\theta\alpha_1 + a_1\vartheta_1^R)}{2(1-\omega_1)} + \vartheta_1^R a_2\right)\frac{(\theta\alpha_1 + a_1\vartheta_1^R)}{\beta^R(1-\omega_1)}$$

$$+\left((1-\theta)\alpha_2 - \frac{\omega_2(\theta\alpha_2 + a_1\vartheta_1^F)}{2(1-\omega_2)} + \vartheta_1^F a_2\right)\frac{(\theta\alpha_2 + a_1\vartheta_1^F)}{\beta^F(1-\omega_2)}$$

$$+\left((1-\theta)\beta_1 - \frac{((1-\theta)\beta_1 + a_2\vartheta_2^R)}{2} + \vartheta_2^R a_2\right)\left(\frac{(1-\theta)\beta_1 + a_2\vartheta_2^R}{\tilde{\beta}^R}\right) \tag{50}$$

$$+\left((1-\theta)\beta_2 - \frac{((1-\theta)\beta_2 + a_2\vartheta_2^F)}{2} + \vartheta_2^F a_2\right)\left(\frac{(1-\theta)\beta_2 + a_2\vartheta_2^F}{\tilde{\beta}^F}\right) > 0.$$

Substituting the results of a_1 and a_2 into Eqs. (37), (38), (41) and (42), and simplifying, we obtain the optimal effort level of hybrid-enabling technological improvements. By substituting optimal values given in Eqs. (48)–(50) into Eqs. (46) and (47) obtain the optimal sharing payoff functions under hybrid-enabling technology on renewable sources and fossil fuel for Player I and Player II.

3.1 The limit of expectation and variance

The payoff of Player I and Player II, under the Stackelberg game paradigm is related to the improvement degree of hybrid-enabling technology via Proposition 4. To analyze the limit of expectations and variance under Stackelberg game equilibrium rewrite (Eq. (19)) as follows.

$$\begin{cases} dK(t) = [\mu_1 + \mu_2 - \xi K(t)]dt + \varphi\sqrt{K}dW(t) \\ K(0) = K_0 > 0, \end{cases} \tag{51}$$

where

$$\mu_1 = \vartheta_1\left[\frac{\alpha_1(2-\theta)(\rho_2+\xi)(\rho_1+\xi) + \vartheta_1^R(\Gamma+\delta)((2-2\theta)(\rho_1+\xi)+\theta(\rho_2+\xi))}{2\beta^R(\rho_2+\xi)(\rho_1+\xi)}\right.$$

$$\left.+\frac{\alpha_2(2-\theta)(\rho_2+\xi)(\rho_1+\xi) + \vartheta_1^F(\Gamma+\delta)((2-2\theta)(\rho_1+\xi)+\theta(\rho_2+\xi))}{2\beta^F(\rho_2+\xi)(\rho_1+\xi)}\right], \tag{52}$$

and

$$\mu_2 = \vartheta_2\left[\frac{(1-\theta)(\beta_1(\rho_2+\xi)+(\Gamma+\delta))\vartheta_2^R}{\tilde{\beta}^R(\rho_2+\xi)} + \frac{(1-\theta)(\beta_2(\rho_2+\xi)+(\Gamma+\delta))\vartheta_2^F}{\tilde{\beta}^F(\rho_2+\xi)}\right]. \tag{53}$$

Proposition 6. *The limit of expectation $E(K(t))$, and variance $D(K(t))$ in the Stackelberg game feedback equilibrium must satisfy*

$$E(K(t)) = \frac{\mu_1+\mu_2}{\xi} + e^{-\xi t}\left(\tilde{K}_0 - \frac{\mu_1+\mu_2}{\xi}\right), \quad \lim_{t\to\infty} E(K(t)) = \frac{\mu_1+\mu_2}{\xi}. \tag{54}$$

$$D(K(t)) = \frac{\varphi^2\left[(\mu_1+\mu_2) - 2\left(\mu_1+\mu_2-\xi\tilde{K}_0\right)e^{-\xi t} + \left(\mu_1+\mu_2-2\xi\tilde{K}_0\right)e^{-2\xi t}\right]}{2\xi^2} \tag{55}$$

$$\lim_{t\to\infty} D(E(K(t))) = \frac{\varphi^2(\mu_1+\mu_2)}{2\xi^2}, \tag{56}$$

Proof. Applying Itô's lemma to (Eq. (51)), obtain:

$$\begin{cases} d(K(t))^2 = \left[2(\mu_1+\mu_2+\varphi^2)K - 2\xi K^2\right]dt + 2\varphi K\sqrt{K}dW(t) \\ (K(0))^2 = \tilde{K}_0^2 > 0. \end{cases} \tag{57}$$

Then $E(K(t))$ and $E(K(t))^2$ can be defined as:

$$\begin{cases} dE(K(t)) = [\mu_1+\mu_2-\xi K(t)]dt \\ K(0) = K_0 > 0. \end{cases} \tag{58}$$

$$\begin{cases} dE(K(t))^2 = \left[[2(\mu_1+\mu_2+\varphi^2)K]E(K) - 2\xi E(K^2)\right]dt \\ (K(0))^2 = K_0^2 > 0, \end{cases} \tag{59}$$

Solving the above non-homogeneous linear differential equation, will obtain the results.

4. Nash non cooperative game

Under Nash-non-cooperative game setting, Player I and Player II simultaneously and independently choose their optimal efforts levels of heterogeneous hybrid-enabling technology sharing concept to maximize their profits.

Proposition 7. *If above conditions are satisfied, the feedback non-cooperative game Nash equilibria will be:*

$$L_N^R = \frac{\theta[\alpha_1(\rho_1+\xi) + (\Gamma+\delta)]}{\beta^R(\rho_1+\xi)}, L_N^F = \frac{\theta[\alpha_2(\rho_1+\xi) + (\Gamma+\delta)]}{\beta^F(\rho_1+\xi)}. \tag{60}$$

$$\tilde{L}_N^R = \frac{(1-\theta)[\beta_1(\rho_2+\xi) + (\Gamma+\delta)]}{\tilde{\beta}^R(\rho_2+\xi)}, \tilde{L}_N^F = \frac{(1-\theta)[\beta_2(\rho_2+\xi) + (\Gamma+\delta)]}{\tilde{\beta}^F(\rho_2+\xi)}, \tag{61}$$

where L_N^R, L_N^F are the optimal level of hybrid-enabling technological advantage on renewable sources and on fossil fuel at time t for Player I, respectively. $\tilde{L}_N^R, \tilde{L}_N^F$ are the optimal level of hybrid-enabling technological advantage on fossil fuel and on renewable sources at time t for Player II, respectively.

The optimal sharing payoff functions under hybrid-enabling technology on renewable sources and on fossil fuel for Player I and Player II are given below

$$V_N^{(I)}(K) = \frac{\theta(\Gamma+\delta_)}{(\rho_1+\xi)}K + \hat{b}_1, \quad V_N^{(II)}(K) = \frac{(1-\theta)(\Gamma+\delta)}{(\rho_2+\xi)}K + \hat{b}_2, \tag{62}$$

where $\hat{b}_1$ and $\hat{b}_2$ are given in the proof.

Proof. See Appendix A.

4.1 The limit of expectation and variance

Proposition 8. *The limit of expectation $E(K(t))$ and variance $D(K(t))$ in the Nash non-cooperative game feedback equilibrium must satisfy*

$$E(K(t)) = \frac{\hat{\mu}_1 + \hat{\mu}_2}{\xi} + e^{-\xi t}\left(K_0 - \frac{\hat{\mu}_1 + \hat{\mu}_2}{\xi}\right), \quad \lim_{t\to\infty} E(K(t)) = \frac{\hat{\mu}_1 + \hat{\mu}_2}{\xi}. \tag{63}$$

$$D(K(t)) = \frac{\varphi^2\left[(\tilde{\mu}_1 + \tilde{\mu}_2) - 2(\tilde{\mu}_1 + \tilde{\mu}_2 - \xi\tilde{K}_0)e^{-\xi t} + (\tilde{\mu}_1 + \tilde{\mu}_2 - 2\xi\tilde{K}_0)e^{-2\xi t}\right]}{2\xi^2} \tag{64}$$

$$\lim_{t\to\infty} D(E(K(t))) = \frac{\varphi^2(\tilde{\mu}_1 + \tilde{\mu}_2)}{2\xi^2}. \tag{65}$$

where $\tilde{\mu}_1 = \frac{\theta\beta^F[\alpha_1(\rho_1+\xi)+(\Gamma+\delta)]+\theta\beta^R[\alpha_2(\rho_1+\xi)+(\Gamma+\delta)]}{\beta^R\beta^F(\rho_1+\xi)}$ *and*
$\tilde{\mu}_2 = \frac{(1-\theta)\beta^F[\beta_1(\rho_2+\xi)+(\Gamma+\delta)]+(1-\theta)\beta^R[\beta_2(\rho_2+\xi)+(\Gamma+\delta)]}{\beta^R\beta^F(\rho_2+\xi)}$.

Poof of Proposition 8 is like the derivation of Proposition 6.

5. Cooperative game

Under cooperative game paradigm, Player I and Player II will choose to collaborate/share their hybrid-enabling technology development knowledge while sharing the payoff function in order to maximize their total payoffs. As a result, hybrid-enabling technology can be improved through this effort as well.

Proposition 9. *If above conditions are satisfied, then the feedback cooperative equilibria are defined as*

$$L_c^R = \frac{(\alpha_1 + \beta_1)(\rho + \xi) + (\Gamma + \delta)(\vartheta_1 + \vartheta_2)}{(\rho + \xi)\beta^R}, \quad L_c^F = \frac{(\alpha_2 + \beta_2)(\rho + \xi) + (\Gamma + \delta)(\vartheta_1 + \vartheta_2)}{(\rho + \xi)\beta^F}, \tag{66}$$

and the optimal cooperative payoff function under hybrid-enabling technology on renewable sources and on fossil fuel, respectively. $V_c(K) = \frac{(\Gamma+\delta)}{(\rho+\xi)}K + \overline{b}$.

where $\overline{b}$, is given in the proof.

Proof. The objective function (optimal sharing payoff function) satisfies the following equation.

$$J(K_0) = \max_{\{L_c^R, L_c^R\}0} E\left\{\int_0^\infty e^{-\rho t}\left[\left(\alpha_1 L_c^R(t) + \alpha_2 L_c^F + \beta_1 \tilde{L}_c^R + \beta_2 \tilde{L}_c^F + (\Gamma + \delta)K\right)\right.\right. \tag{67}$$

Then the optimal revenue sharing function satisfies the following HJB equation

$$\begin{aligned}\rho V_c(K) = \max_{\{L_c^R, L_c^F\}\geq 0} &\left\{\left[\alpha_1 L_c^R(t) + \alpha_2 L_c^F + \beta_1 \tilde{L}_c^R + \beta_2 \tilde{L}_c^F + (\Gamma + \delta)K)\right]\right. \\ &-\frac{1}{2}\beta^R\left(L_c^R\right)^2 - \frac{1}{2}\beta^F\left(L_c^F\right)^2 \\ &\left.+\frac{\partial V_c(K)}{\partial K}\left[\vartheta_1\left(L_c^R, L_c^F\right) + \vartheta_2\left(\tilde{L}_c^R, \tilde{L}_c^F\right) - \xi K\right] + \frac{1}{2}\frac{\partial^2 V_c(K)}{\partial K^2}\varphi^2(K)\right\}\end{aligned} \tag{68}$$

Via the first order conditions, now obtain the optimal values $\left(L_c^R, L_c^F\right)$ as:

$$L_c^R = \frac{(\alpha_1 + \beta_1) + V'_c(K)(\vartheta_1 + \vartheta_2)}{\beta^R}, \tag{69}$$

$$L_c^F = \frac{(\alpha_2 + \beta_2) + V'_c(K)(\vartheta_1 + \vartheta_2)}{\beta^F}. \tag{70}$$

Substituting the results of Eqs. (69) and (70), obtain

$$\begin{aligned} \rho V_c(K) = & \max_{\{L_c^R, L_c^F\} \geq 0} \left\{ (\alpha_1 + \beta_1)\left(\frac{(\alpha_1 + \beta_1) + V'_c(K)(\vartheta_1 + \vartheta_2)}{\beta^R}\right) \right. \\ & + (\alpha_2 + \beta_2)\left(\frac{(\alpha_2 + \beta_2) + V'_c(K)(\vartheta_1 + \vartheta_2)}{\beta^F}\right) - \frac{1}{2}\beta^R\left(\frac{(\alpha_1 + \beta_1) + V'_c(K)(\vartheta_1 + \vartheta_2)}{\beta^R}\right)^2 \\ & - \frac{1}{2}\beta^F\left(\frac{(\alpha_2 + \beta_2) + V'_c(K)(\vartheta_1 + \vartheta_2)}{\beta^F}\right)^2 - \frac{\partial V_c(K)}{\partial K}\xi K \\ & + \frac{\partial V_c(K)}{\partial K}\left[(\vartheta_1 + \vartheta_2)\left(\frac{(\alpha_1 + \beta_1) + V'_c(K)(\vartheta_1 + \vartheta_2)}{\beta^R}\right)\right. \end{aligned} \tag{71}$$

Hence, the solution of the HJB equation is an unary function with K, $V_c = \overline{a}K + \overline{b}$, where $\overline{a}$ and $\overline{b}$ are constant that need to be solved. This implies that

$$\overline{a} = \frac{(\Gamma + \delta)}{(\rho + \xi)}. \tag{72}$$

$$\begin{aligned} \overline{b} = & \left((\alpha_1 + \beta_1) - \frac{((\alpha_1 + \beta_1) + \overline{a}(\vartheta_1 + \vartheta_2))}{2} + (\vartheta_1 + \vartheta_2)\overline{a}\right)\left(\frac{(\alpha_1 + \beta_1) + \overline{a}(\vartheta_1 + \vartheta_2)}{\beta^R}\right) \\ & + \left((\alpha_2 + \beta_2) - \frac{((\alpha_2 + \beta_2) + \overline{a}(\vartheta_1 + \vartheta_2))}{2} + (\vartheta_1 + \vartheta_2)\overline{a}\right)\left(\frac{(\alpha_2 + \beta_2) + \overline{a}(\vartheta_1 + \vartheta_2)}{\beta^F}\right) > 0. \end{aligned} \tag{73}$$

Substituting the results of Eqs. (72) and (73), into $V_c = \overline{a}K + \overline{b}$, will obtain the results.

5.1 The limit of expectation and variance

Proposition 10. The limit of expectation and variance in cooperative game feedback equilibrium satisfy

$$E(K(t)) = \frac{\overline{\mu}_1 + \overline{\mu}_2}{\xi} + e^{-\xi t}\left(\tilde{K}_0 - \frac{\overline{\mu}_1 + \overline{\mu}_2}{\xi}\right), \qquad \lim_{t \to \infty} E(K(t)) = \frac{\overline{\mu}_1 + \overline{\mu}_2}{\xi}. \tag{74}$$

$$D(K(t)) = \frac{\varphi^2\left[(\overline{\mu}_1 + \overline{\mu}_2) - 2(\overline{\mu}_1 + \overline{\mu}_2 - \xi K_0)e^{-\xi t} + (\overline{\mu}_1 + \overline{\mu}_2 - 2\xi K_0)e^{-2\xi t}\right]}{2\xi^2} \tag{75}$$

$$\lim_{t \to \infty} D(E(K(t))) = \frac{\varphi^2(\overline{\mu}_1 + \overline{\mu}_2)}{2\xi^2}, \tag{76}$$

where $\overline{\mu}_1 = \frac{(\alpha_1 + \beta_1)(\rho + \xi) + (\Gamma + \delta)(\vartheta_1 + \vartheta_2)}{(\rho + \xi)}$ and $\overline{\mu}_2 = \frac{(\alpha_2 + \beta_2)(\rho + \xi) + (\Gamma + \delta)(\vartheta_1 + \vartheta_2)}{(\rho + \xi)}$.

Proof. Poof of Proposition 10 is like the derivation of Propositions 6 and 8.

6. Comparative analysis of equilibrium results

Proposition 11. *The outcome of the game depends on the parameters of the game and the type of the equilibrium one considers.*

Proof. (i) Player I, will participate in a Stackelberg game to share more hybrid-enabling technology under the condition that Player II pay much more extra cost for hybrid-enabling technology

$$\frac{\alpha_1(2-\theta)(\rho_2+\xi)(\rho_1+\xi)+\vartheta_1^R(\Gamma+\delta)((2-2\theta)(\rho_1+\xi)+\theta(\rho_2+\xi))}{2\beta^R(\rho_2+\xi)(\rho_1+\xi)} - \frac{(1-\theta)(\beta_1(\rho_2+\xi)+(\Gamma+\delta))\vartheta_2^R}{\tilde{\beta}^R(\rho_2+\xi)} > 0, \tag{77}$$

and

$$\frac{\alpha_2(2-\theta)(\rho_2+\xi)(\rho_1+\xi)+\vartheta_1^F(\Gamma+\delta)((2-2\theta)(\rho_1+\xi)+\theta(\rho_2+\xi))}{2\beta^F(\rho_2+\xi)(\rho_1+\xi)} - \frac{(1-\theta)(\beta_2(\rho_2+\xi)+(\Gamma+\delta))\vartheta_2^F}{\tilde{\beta}^F(\rho_2+\xi)} > 0. \tag{78}$$

(ii) Player I will prefer to participate in a cooperative game over a non-cooperative game with Player II under the condition such that

$$\frac{(\alpha_1+\beta_1)(\rho+\xi)+(\Gamma+\delta)(\vartheta_1+\vartheta_2)}{(\rho+\xi)\beta^R} - \frac{\theta[\alpha_1(\rho_1+\xi)+(\Gamma+\delta)]}{\beta^R(\rho_1+\xi)} > 0, \tag{79}$$

and

$$\frac{(\alpha_2+\beta_2)(\rho+\xi)+(\Gamma+\delta)(\vartheta_1+\vartheta_2)}{(\rho+\xi)\beta^F} - \frac{\theta[\alpha_2(\rho_1+\xi)+(\Gamma+\delta)]}{\beta^F(\rho_1+\xi)} > 0. \tag{80}$$

(iii) The total payoff for Player I under a Stackelberg game exceeds the total payoff of Nash non-cooperative game with Player II under the condition such that

$$\frac{\alpha_1(2-\theta)(\rho_2+\xi)(\rho_1+\xi)+\vartheta_1^R(\Gamma+\delta)((2-2\theta)(\rho_1+\xi)+\theta(\rho_2+\xi))}{2\beta^R(\rho_2+\xi)(\rho_1+\xi)} - \frac{(\alpha_1+\beta_1)(\rho+\xi)+(\Gamma+\delta)(\vartheta_1+\vartheta_2)}{(\rho+\xi)\beta^R} > 0, \tag{81}$$

and

$$\frac{\alpha_2(2-\theta)(\rho_2+\xi)(\rho_1+\xi)+\vartheta_1^F(\Gamma+\delta)((2-2\theta)(\rho_1+\xi)+\theta(\rho_2+\xi))}{2\beta^F(\rho_2+\xi)(\rho_1+\xi)} - \frac{(\alpha_2+\beta_2)(\rho+\xi)+(\Gamma+\delta)(\vartheta_1+\vartheta_2)}{(\rho+\xi)\beta^F} > 0. \tag{82}$$

(iv). Player II will prefer to participate in a cooperative game over a non-cooperative game with Player I under the condition such that

$$\frac{(\alpha_1+\beta_1)(\rho+\xi)+(\Gamma+\delta)(\vartheta_1+\vartheta_2)}{(\rho+\xi)\beta^R}-\frac{(1-\theta)[\beta_1(\rho_2+\xi)+(\Gamma+\delta)]}{\tilde{\beta}^R(\rho_2+\xi)}>0, \quad (83)$$

and

$$\frac{(\alpha_2+\beta_2)(\rho+\xi)+(\Gamma+\delta)(\vartheta_1+\vartheta_2)}{(\rho+\xi)\beta^F}-\frac{(1-\theta)[\beta_2(\rho_2+\xi)+(\Gamma+\delta)]}{\tilde{\beta}^F(\rho_2+\xi)}>0. \quad (84)$$

Proposition 12. *For any $K\geq 0$, under the condition that Player II pay an extra cost for sharing hybrid-enabling technology. Then the optimal sharing payoff of hybrid-enabling technology of Player I reaches higher than the optimal sharing payoff under the condition that player II does not provide extra cost. This implies that $V_S^{(\mathrm{I})}(K)\geq V_N^{(\mathrm{I})}(K)$. Similarly, the optimal sharing payoff of hybrid-enabling technology of Player II reaches higher than the optimal sharing payoff under the condition that Player II do not provide extra cost, such that $V_S^{(\mathrm{II})}(K)\geq V_N^{(\mathrm{II})}(K)$.*

Proof. When $0\leq\theta\leq\frac{2}{3}$, establish that

$$\Delta V^{(\mathrm{I})}(K)=V_S^{(\mathrm{I})}(K)-V_N^{(\mathrm{I})}(K)=\frac{\theta(\Gamma+\delta)}{(\rho_1+\xi)}K+b_1-\frac{\theta(\Gamma+\delta)}{(\rho_1+\xi)}K+\hat{b}_1$$
$$=b_1-\hat{b}_1>0, \quad (85)$$

and

$$\Delta V^{(\mathrm{II})}(K)=V_S^{(\mathrm{II})}(K)-V_N^{(\mathrm{II})}(K)=\frac{(1-\theta)(\Gamma+\delta_))}{(\rho_1+\xi)}K+b_2-\frac{(1-\theta)(\Gamma+\delta)}{(\rho_1+\xi)}K+\hat{b}_2$$
$$=b_2-\hat{b}_2>0. \quad (86)$$

7. Concluding remarks

In this chapter a complete study of an energy market by considering a Bertrand duopoly game with two power plants using endogenous hybrid-enabling technology was presented. Numerous game paradigms were articulated and defined including Stackelberg, Nash non-cooperative and cooperative games as well as their relevant equilibria via a feedback control strategy. Mathematically, the necessary conditions under which a power plant will move from taking part in a non-cooperative Nash game to participate as a leader in a Stackelberg game was derived. In doing so, this model allowed us to quantify the optimal level of subsidy for sharing the hybrid-enabling technology. We then adopted the concept of limit expectation and variance of the improvement degree to identify the influence of random factors of external environment and limitations of the decision maker. It is found that for a given level of payoff distribution the Stackelberg equilibria with technological enhancements, the knowledge sharing paradigm dominates the Nash equilibria. In both Stackelberg and Nash games, optimal technological enhancements for power plants were found to be proportional to the government subsidy, but the variance improvement degree of the Stackelberg game differed to the results of the Nash non-cooperative game due to the influence of random factors.

Furthermore, we have shown that due to optimal price reaction functions being upward sloping, the subsidy level plays a decisive role on the payoff function of power plant II as the leader in a Stackelberg game. This model shows that cost reducing R&D investments with efficient hybrid-enabling technology innovation/s strengthens one's competitive bargaining position via the level of subsidy for Power Plant I to become a follower in the Stackelberg game. By analyzing this stochastic differential game model, we capture the government subsidy incentive as well as the subsidy that the leader (Power Plant II) pays the follower (Power Plant I) to share hybrid-enabling technology.

The proposed quantitative framework could assist policymakers when determining the appropriate R&D incentives for the development of hybrid-enabling technology within the energy market to achieve desired short and long-term environmental objectives with respect to budget limitations and environmental considerations.

A. Appendix

Proof. The optimal profit function for power plant I satisfies the following HJB equation such that $V_N^{(I)}(K)$:

$$\begin{aligned}\rho_1 V_N^{(I)}(K) = \max_{\{L_N^R, L_N^F\} \geq 0} & \Big\{ \Big[\theta \Big(\alpha_1 L_N^R(t) + \alpha_2 L_N^F + \beta_1 \tilde{L}_N^R + \beta_2 \tilde{L}_N^F + (\Gamma + \delta) K \Big) \Big] \\ & - \frac{1}{2} \beta^R \left(L_N^R \right)^2 - \frac{1}{2} \beta^F \left(L_N^F \right)^2 \\ & + \frac{\partial V_N^{(I)}(K)}{\partial K} \Big[\vartheta_1 \left(L_N^R, L_N^F \right) + \vartheta_2 \left(\tilde{L}_N^R, \tilde{L}_N^F \right) - \xi K \Big] + \frac{1}{2} \frac{\partial^2 V_N^{(I)}(K)}{\partial K^2} \varphi^2(K) \Big\}\end{aligned} \tag{A.1}$$

Via the first order conditions, first obtain the optimal values $\left(L_N^R, L_{(I)}^R \right)$ for power plant I as:

$$L_N^R = \frac{\theta \alpha_1 + V'^{(I)}_N(K) \vartheta_1^R}{\beta^R} = \frac{\theta \alpha_1 + a_1 \vartheta_1^R}{\beta^R}, \tag{A.2}$$

$$L_N^F = \frac{\theta \alpha_2 + V'^{(I)}_N(K) \vartheta_1^F}{\beta^F} = \frac{\theta \alpha_2 + a_1 \vartheta_1^F}{\beta^F} \tag{A.3}$$

where $\frac{\partial V_N^{(I)}(K)}{\partial K} \equiv V'^{(I)}_N(K)$. The HJB for power plant (II), using $\frac{\partial V_N^{(II)}(K)}{\partial K} = V'^{(II)}_N(K)$, then obtain

$$\begin{aligned}\rho_2 V_N^{(II)}(K) = \max_{\{\tilde{L}_N^R, \tilde{L}_N^F\}} & \Big\{ \Big[(1-\theta) \Big(\alpha_1 L_N^R(t) + \alpha_2 L_N^F + \beta_1 \tilde{L}_N^R + \beta_2 \tilde{L}_N^F + (\Gamma + \delta) K \Big) \Big] \\ & - \frac{1}{2} \tilde{\beta}^R \left(\tilde{L}_N^R \right)^2 - \frac{1}{2} \tilde{\beta}^F \left(\tilde{L}_N^F \right)^2 \\ & + \frac{\partial V_N^{(I)}(K)}{\partial K} \Big[\vartheta_1 \left(L_N^R, L_N^F \right) + \vartheta_2 \left(\tilde{L}_N^R, \tilde{L}_N^F \right) - \xi K \Big] + \frac{1}{2} \frac{\partial^2 V_N^{(I)}(K)}{\partial K^2} \varphi^2(K) \Big\}.\end{aligned} \tag{A.4}$$

Substituting Eqs. (A.2) and (A.3) results to Eq. (A.4) and via the first order conditions, we obtain the optimal values $\left(\tilde{L}_N^R, \tilde{L}_N^F \right)$ for power plant II as:

$$\tilde{L}_N^R = \frac{(1-\theta)\beta_1 + V'^{(II)}_N(K)\vartheta_2^R}{\tilde{\beta}^R} = \frac{(1-\theta)\beta_1 + a_2\vartheta_2^R}{\tilde{\beta}^R}, \tag{A.5}$$

$$\tilde{L}_N^F = \frac{(1-\theta)\beta_2 + V'^{(II)}_N(K)\vartheta_2^F}{\tilde{\beta}^F} = \frac{(1-\theta)\beta_2 + a_2\vartheta_2^F}{\tilde{\beta}^F}. \tag{A.6}$$

Hence, the solution of the HJB equation is an unary function with K, such that $V_N^{(I)} = \hat{a}_1 K + \hat{b}_1$, $V_N^{(II)} = \hat{a}_2 K + \hat{b}_2$. Hence, finally $\hat{a}_1, \hat{b}_1, \hat{a}_2, \text{and} \hat{b}_2$ as:

$$\hat{a}_1 = \frac{\theta(\Gamma+\delta)}{(\rho_1+\xi)}, \quad \hat{b}_1 = \frac{\hat{\Phi}_1}{\rho_1}, \quad \hat{a}_2 = \frac{(1-\theta)(\Gamma+\delta)}{(\rho_2+\xi)}, \quad \hat{b}_2 = \frac{\hat{\Phi}_2}{\rho_2}, \tag{A.7}$$

where

$$\begin{aligned}\hat{\Phi}_1 = &\left(\alpha_1\theta - \frac{(\theta\alpha_1 + a_1\vartheta_1^R)}{2} + \vartheta_1^R a_1\right)\left(\frac{\theta\alpha_1 + a_1\vartheta_1^R}{\beta^R}\right)\\ &+\left(\alpha_2\theta - \frac{(\theta\alpha_2 + a_1\vartheta_1^F)}{2} + \vartheta_1^F a_1\right)\left(\frac{\theta\alpha_2 + a_1\vartheta_1^F}{\beta^F(1-\omega_2)}\right)\\ &+(\beta_1\theta + \vartheta_2^R a_1)\left(\frac{(1-\theta)\beta_1 + a_2\vartheta_2^R}{\tilde{\beta}^R}\right) + (\beta_2\theta + \vartheta_2^F a_1)\left(\frac{(1-\theta)\beta_2 + a_2\vartheta_2^F}{\tilde{\beta}^F}\right) > 0.\end{aligned} \tag{A.8}$$

and

$$\begin{aligned}\hat{\Phi}_2 = &((1-\theta)\alpha_1 + \vartheta_1^R a_2)\frac{(\theta\alpha_1 + a_1\vartheta_1^R)}{\beta^R(1-\omega_1)}\\ &+((1-\theta)\alpha_2 + \vartheta_1^F a_2)\frac{(\theta\alpha_2 + a_1\vartheta_1^F)}{\beta^F(1-\omega_2)}\\ &+\left((1-\theta)\beta_1 - \frac{((1-\theta)\beta_1 + a_2\vartheta_2^R)}{2} + \vartheta_2^R a_2\right)\left(\frac{(1-\theta)\beta_1 + a_2\vartheta_2^R}{\tilde{\beta}^R}\right)\\ &+\left((1-\theta)\beta_2 - \frac{((1-\theta)\beta_2 + a_2\vartheta_2^F)}{2} + \vartheta_2^F a_2\right)\left(\frac{(1-\theta)\beta_2 + a_2\vartheta_2^F}{\tilde{\beta}^F}\right) > 0.\end{aligned} \tag{A.9}$$

Substituting $a_1 \& a_2$ into Eqs. (A.2), (A.3), (A.5) and (A.6) and simplifying, obtain the optima effort level of technological improvements. By substituting the above results into Eqs. (A.1) and (A.4) we obtain the optimal sharing payoff functions, for power plant (I) and power plant (II).

Author details

Ryle S. Perera
Macquarie University Business School, Macquarie University, Sydney, NSW, Australia

*Address all correspondence to: ryle.perera@mq.edu.au

References

[1] Khan SAR, Sharif A, Golpira H, Kumar A. A green ideology in Asian emerging economies: From environmental policy and sustainable development. Sustainable development. 2019;**27**:1063-1075

[2] Khan SAR, YU Z, Sharif A, Golpira H. Determinants of economic growth and environmental sustainability in south Asian Association for Reginal Cooperation: Evidence from panel ARDL. Environmental Science and pollution Research. 2000. DOI: https://doi.org/10.1007/s11356-020-10410-1

[3] Khan SAR, YU Z, Kumar A, Zavadskas E. Measuring the impact of renewable energy, public heath expenditure, logistics, and environmental performance on sustainable economic growth. *Sustainable* development. 2020;**28**:833-843

[4] Khan SAR, YU Z, Belhadi A, Mardani A. Investigating the effects of renewable energy on international trade and environmental quality. Journal of Environmental management. 2020. DOI: https://doi.org/10.1016/j.jenvman.2020.11089

[5] Khan SAR, YU Z, Golpira H, Sharif A, Mardani A. A state-of-the-art review and meta-analysis on sustainable supply chain management: Future research directions. Journal of Cleaner Production. mental management. 2020. DOI: https://doi.org/10.1016/j.jenvman.2020.123357

[6] Vespucci MT, Allevi E, Gnudi A, Innorta M. Cournot equilibria in oligopolistic electricity markets. IMA Journal Management Mathematics. 2010;**21**:183-193

[7] Li T, Shahidehpour M, Keyhani A. Market power analysis in electricity markets using supply function equilibrium model. IMA Journal of Management Mathematics. 2004;**15**(4): 339-354

[8] Hinz J. Equilibrium strategies in random-demand procurement auctions with sunk costs. IMA Journal of Management Mathematics. 2006;**17**(1): 61-81

[9] Mahmoudi R, Hafezalkotob A, Makui A. Source selection problem of competitive power plants under government intervention: A game theory approach. Journal of Industrial Engineering International. 2014;**10**(3): 1-15

[10] Dinan TM. Economic efficiency effects of alternative policies for reducing waste disposal. Journal of Environmental Economic Management. 1993;**25**:242-256

[11] Dobbs IM. Litter and waste management: Disposal taxes versus user chargers. Canadian Journal of Economics. 1991;**24**:221-227

[12] Menniti D, Pinnarelli A, Sorrentino N. Simulation of producers behavior in the electricity market by evolutionary games. Electrical Power System Research. 2008; **87**:265-281

[13] Ulph A. Environmental policy and international trade when governments and producers act strategically. Journal of Environmental Economic Management. 1996;**30**:242-256

[14] Webb JN. *Game Theory Decisions, Interaction and Evolution*. London: Springer; 2007

[15] Perera RS. An evolutionary game theory strategy for carbon emission reduction in the electricity market. International Game Theory Review. 2018;**20**(4):1-20

[16] Perera RS. Trans-boundary emission under stochastic differential game. International Game Theory Review. 2020 https://doi.org/10.1142/S0219198920500097

[17] Isaacs R. *Differential games: A mathematical theory with applications to warfare and pursuit*. In: *Control and Optimization*. New York: Wiley; 1965

[18] Maynard Smith J. *Evolution and the Theory of Games*. Cambridge University Press; 1974

[19] McKelvey RD, Palfrey TA. Quantal response equilibria for normal form games. Games Economic Behavior. 1995;**10**:6-38

[20] Riechmann T. Genetic algorithm learning and evolutionarily games. Journal of Economic Dynamic Control. 2001;**25**(6–7):1019-1037

[21] Fleming WH, Souganidis PE. On the existence of value function of two-player, zero-sum stochastic differential games. Indiana University Mathematics Journal. 1989;**38**(2):293-314

[22] Kyoto Protocol. (1992). http://unfccc.int/kyoto_protocol/items/2830.php, 11 December 1997

[23] Vives X. On the efficiency of Bertrand and Cournot equilibia with product differentiation. Journal Economic Theory. 1985;**36**:166-175

[24] Friedman D. Evolutionary games in economics. Econometrica. 1991;**59**(3):637-666

[25] Weibull JW. *Evolutionary Game Theory*. MIT Press; 1997

[26] Xiao T, Chen G. Wholesale pricing and evolutionarily stable strategies of retailers with imperfectly observable objective. European Journal of Operational Research. 2009;**196**(3):1190-1201

[27] Nash JF. Equilibrium points in n-person games. Proceedings of the National Academy of Sciences of the United States of America. 1950;**36**:48-49

[28] Nash JF. Non-cooperative games. Annals of Mathematics. 1951;**54**:286-295

[29] Brander JA, Spencer BJ. Strategic commitments with R&D: The symmetric case. Bell Journal of Economics. 1983;**14**(1):225-235

[30] Brander JA, Lewis TR. Oligopoly and financial structure: The limited liability effect. American Economic Review. 1986;**76**(5):956-970

[31] Dixit A. The role of investment entry-deterrence. The Economic Journal. 1980;**90**:95-106

[32] Gal-Or E. First mover and second mover advantages. International Economic Review. 1985;**26**:649-653

[33] Milgrom P, Roberts J. Rationalizability, learning and equilibrium in games with strategic complementarities. Econometrica. 1990;**58**:1255-1277

[34] Van Long N. *A Survey of Dynamic Games in Economics*. World Scientific Publishing. Singapore; 2010

[35] Colman AM, Bacharach M. Payoff dominance and the Stacklberg heuristic. Theory and Decision. 1997;**43**:1), 1-1),19

[36] Von Neumann J, Morgenstern O. *Theory of Games and Economic Behavior*. Princeton: Princeton University Press; 1944

Chapter 6

Eggshell and Seashells Biomaterials Sorbent for Carbon Dioxide Capture

Abarasi Hart and Helen Onyeaka

Abstract

This review aims to explore the application of natural and renewable bioceramics such as eggshell and seashells in carbon dioxide (CO_2) capture from power plant flue gas. CO_2 capture, utilisation and storage (CCUS) is considered a means to deliver low carbon energy, decarbonising industries, power plants and facilitates the net removal of CO_2 from the atmosphere. The stages involved include CO_2 capture, transport of the captured CO_2, utilisation and secure storage of the captured CO_2. This chapter reports the use of eggshell and seashells biomaterials as an adsorbent to separate CO_2 from other gases generated by power plants and industrial processes. The capture of carbon dioxide by adsorption is based on the ability of a material to preferentially adsorb or carbonate CO_2 over other gases. In light of this, calcined eggshell and seashells biomaterial rich in calcium carbonate from which calcium oxide (94%) can be obtained have demonstrated a strong affinity for CO_2. These biomaterials are abundant and low-cost alternative to zeolite, activated carbon and molecular sieve carbon. The mechanism of CO_2 capture by eggshell and seashells derived CaO adsorbent comprises of a series of carbonation-calcination reactions (CCR): calcium oxide (CaO) reacts with CO_2 resulting in calcium carbonate ($CaCO_3$), which releases pure CO_2 stream upon calcinations for sequestration or utilisation, and as a consequence, the biomaterial is regenerated. Findings reveal that these biomaterials can hold up to eight times its own weight of CO_2 from flue gas stream. It was also found that the combination of 2 M acetic acid and water pretreatment improved the reactivity and capture capacity of the biomaterial for successive regeneration over four cycle's usage. Unlike activated carbon, these biomaterials are considered stable for high-temperature adsorption through carbonation.

Keywords: carbon dioxide capture, eggshell, seashells, adsorption, carbonation, calcination

1. Introduction

The combustion of fossil fuels such as coal, oil and natural gas for energy generate a large amount of carbon dioxide (CO_2) emission, causing global warming and climate change. Presently, legislation such as the Paris Agreement of 2015, provided a framework on dealing with greenhouse-gas-emissions (GHG) mitigation, and it is anticipated across the industrialised world to cut down the amount of CO_2

emissions and limit global warming to less than 2°C [1]. Additionally, the demand for energy is expected to increase by 50% in 2030, and also oil and gas are considered the principal feedstock of about 90% of chemicals produced worldwide, and it is forecasted that petrochemical industries will become the largest driver for global oil consumption by 2050 [2]. In this light, it is therefore important to mitigate the environmental impact of burning carbon-based fuels, in which potential progress has already been made in CO_2 capture, utilisation and storage (CCUS) technologies [3]. The CCUS is considered a means to deliver low carbon energy, decarbonising industries, and facilitates the net removal of CO_2 from the atmosphere. The stages involved include CO_2 capture, transport of the captured CO_2, utilisation and secure storage of the captured CO_2.

Carbon dioxide capture will play a significant role as fossil fuel will continue to meet world energy needs during this transition to sustainable low-carbon energy system [4]. It has also been reported that this transition phase will linger for a long time, providing sufficient time for the development and commercialisation of renewable energy systems. The transportation sector especially logistics operations majorly depend on fossil fuels, resulting in large carbon footprint on the environment. Based on World Bank data, the shift into low-carbon energy such as renewable energy in logistics operations prove to minimise carbon emission and other greenhouse gases, create sustainable environment as well as improve economic performance [5, 6]. In 2018, the global CO_2 emissions increased to 37.1 Gt which is forecasted to rise by about 10% in 2040, majorly due to the combustion of fossil fuels from industrial processes and transportation sector [7]. Hence, the impact of carbon emissions from logistics operations on the environment, global warming, climate change and health can be reduced remarkably by adopting renewable energy and green vehicles [6]. Therefore, government policy and legislations such as the Paris Agreement of 2015 are necessary to drive research and development into low-carbon energy and environmental sustainability. As a result of these policies, renewable energy and carbon capture technologies are being developed, and their implementation is expected to improve environmental quality and sustainability [5, 8, 9]. Unlike fossil fuels, renewable energies promote eco-friendly environment. Hence, CCUS technologies will enable the use of fossil fuels in a cleaner way when integrated with power plants to mitigate global warming and climate change effects. CO_2 has found utilisation in the following areas mineralisation, biological utilisation, food and beverages, energy storage media, chemicals, enhanced oil recovery, coal bed methane and hydraulic fracturing processes [7]. However, public awareness and acceptance of CCUS is still low in spite of the attention shown by the scientific communities, industries and governments. Findings by Tcvetkov et al. [10] show that most studies on CCUS are dedicated to carbon dioxide storage in geological formation with less attention on capture and transportation. Hence, this study focuses on carbon dioxide capture using natural and renewable biomaterials such as eggshells and seashells.

The essence of carbon capture is to separate carbon dioxide from other gases produced as a result of the combustion of fossil fuels for power generation and industrial processes. **Figure 1** shows the three main approaches to accomplish this, which are pre-combustion capture, post-combustion capture and oxy-fuel combustion methods.

Before now, the capture of carbon dioxide is commonly achieved in the industry through absorption using liquid solvents such as selexol, rectisol, and mono-ethanol-amine, MEA [11]. The absorption process involves the use of two columns, namely the absorber and the stripper. This makes the process cost intensive in addition to corrosion issues. Consequently, a large amount of energy is needed to absorb CO_2 [12]. On the other hand, physical adsorption via solid adsorption

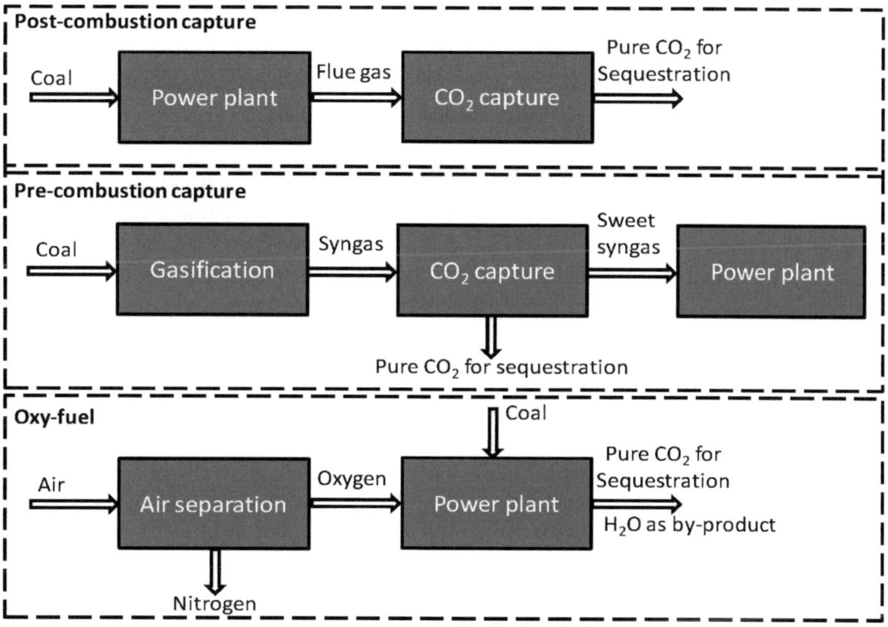

Figure 1.
Three common carbon dioxide capture approaches for coal fired power plant.

processes can selectively separate carbon dioxide from flue gas mixture. The advantages of adsorption include high selectivity, operation simplicity, low-cost, ease of regeneration, and low corrosiveness of adsorbent compared with solvent processes [11, 12].

The carbon dioxide adsorption approaches rely on the ability of the adsorbing material to preferentially adsorb CO_2 over other gases. This is achieved through a packed bed system of the adsorbent materials. The adsorbent materials will continue to absorb CO_2 until it is saturated, which is its adsorptive capacity. At this point, the packed bed undergoes desorption either through pressure swing adsorption (PSA) or temperature swing adsorption (TSA), which causes the release of the adsorbed CO_2 to the point where the adsorbent material is at equilibrium [12, 13]. The commonly used adsorbent materials include zeolites, activated carbon, microporous/mesoporous silica, carbonates, carbon molecular sieves and metal organic frameworks. These materials possess adequate surface area and pore network structures that are highly microporous to accommodate and capture CO_2 [12, 13]. The adsorbent materials are evaluated on the basis of adsorption capacity, preferential adsorption affinity for carbon dioxide over gases from flue gas stream, adsorption and desorption kinetics, low-cost, tolerance of impurities, mechanical strength, multicycle durability and regeneration of stability [13]. Additionally, the porous structure of the adsorbent material is engineered to improve mass transport by reducing diffusional resistance, and the microstructure and morphological texture must demonstrate the capacity to hold captured CO_2 during multi cycling between the absorption and regeneration steps [13, 14]. However, since the process is based on gas-solid interaction, operational conditions such as gas flow rate, temperature and vibration could cause disintegration of adsorbent material due to crushing and abrasion, and consequently collapse pore network structures. It is also rare to find a single adsorbent material that maximises all the above highlighted attributes. Therefore, this review explores the use of other materials such as eggshell and seashell rich in calcium carbonate through reactive adsorption, which involves carbonation – calcination of $CaO/CaCO_3$ for carbon dioxide capture.

Alkaline earth metal oxides have demonstrated a strong affinity for acidic gas such as carbon dioxide and sulphur oxides. These metal oxides, particularly calcium oxide (CaO), are effective for the removal of CO_2 via carbonation at moderate temperatures of less than 700°C [11]. Hence, calcium oxide has proven a good sorbent material for carbon dioxide capture. With regards to availability and cost, an excellent source of CaO is calcium carbonate ($CaCO_3$). The most widely natural source of $CaCO_3$ includes dolomite and limestone. However, these natural resources are non-renewable, energy intensive to exploit, their mining cause damage to the environment as well as landscape. More also, CaO sorbent derived from natural limestone decreases in its reactivity over a number of cycles of reaction with CO_2 [15]. As a result of this, attention has been shifted to renewable sources such as eggshells, seashells and snail shells. These waste biomaterials provide sustainable source of calcium carbonate ($CaCO_3$) in the range of 90–96% [16]. Calcined eggshell and seashells such as oyster shell are rich in lime (CaO) and can be combined with post-combustion and pre-combustion systems to separate CO_2 through cyclic carbonation of CaO (calcined eggshell/seashell) to $CaCO_3$, and subsequently the calcination of $CaCO_3$ to release pure CO_2 and regenerate back to CaO, as shown in **Figure 2** [15, 17–19]. This reversible reaction between CaO and CO_2 is a promising approach of removing CO_2 from flue gas from power plants, producing a pure stream of CO_2 ready for geological sequestration [15, 19]. To achieve this objective, the material should exhibit sufficient reactivity and thermal stability. Eggshell and seashell are a low-cost and abundant alternative to synthetic calcium carbonate and lime sorbents.

The poultry and seafood industries generate millions of tonnes of waste shells annually, which are disposed of in landfills. These biomaterials are rich in calcium carbonate, and subsequently, a large source of calcium oxide. The discarded eggshells and seashells after consumption of their food content, the heap waste shell is a habitat for microbes which causes environmental and air pollution due to emission of intensive odour especially during microbial decomposition [16]. These waste shell biomaterials can be recycled and used as a source of calcium oxide material for carbon dioxide capture purposes. Remarkable costs can be saved when these waste shells biomaterials are re-used, with emphasises on economic and sustainable

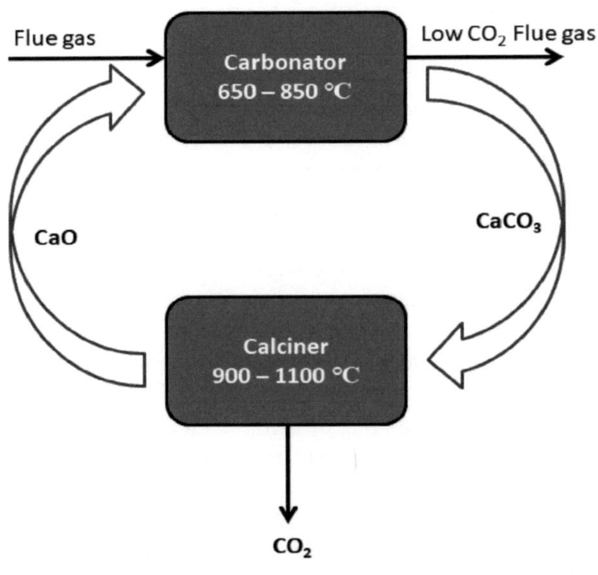

Figure 2.
Carbonation – calcination process in calcium looping cycle application for carbon capture.

environmental benefits of recycling instead of disposing. However, the carbon dioxide capture capacity of synthesised calcium oxide sorbents from eggshell and seashells decreases, as cycles of carbonation and calcination increases because of sintering over time [17]. To remedy this, it is important to generate more porous surface structure in the biomaterials through pre-treatment and regeneration processes.

2. Physicochemical properties of eggshell and seashells biomaterials

The major solid mineral component of eggshells and seashells is calcium carbonate in the range of 92–96% and minor trace elements such as silica, alumina, phosphorous, magnesium, sodium, potassium, zinc, manganese, iron, and copper. A detail composition of eggshell and seashells has been reported elsewhere [16]. The physical properties of some calcined eggshells and seashells biomaterials such as surface area, pore volume and pore diameter are shown in **Table 1**. These waste shells biomaterials exhibit the type-IV isotherm which an attribute of mesoporous texture morphology characterised with a network of micropores. The pore size re-affirms their microstructure characteristics to accommodate captured CO_2. During calcination, the specific surface area and pore volume of the crushed eggshells and seashells biomaterials increases, as the calcination temperature increases. This is because of the evolution of porosity within the material as a result of the release of CO_2 from $CaCO_3$, leading to the formation of CaO [16, 20]. However, at a temperature greater than 900°C, the surface area and pore volume decreased due to prolonged thermal effect, resulting in sintering [16, 20].

Figure 3 shows the X-Ray Diffraction (XRD) patterns of uncalcined (natural) and calcined (thermally treated) eggshell (quail) and seashell (oyster shell). The major component visible on the XRD pattern of the natural crushed shells is $CaCO_3$

Parameter	Mussel shell	Oyster shell	Chicken eggshell	Ostrich eggshell
Surface area (m^2/g)	89.91	24.00	54.60	71.00
Pore volume (cm^3/g)	0.130	0.04	0.015	0.022
Pore size (nm)	3.5	6.6	0.54	0.61
Reference	[21]		[22]	

Table 1.
Surface area, pore size and volume of calcined seashells and eggshells biomaterials.

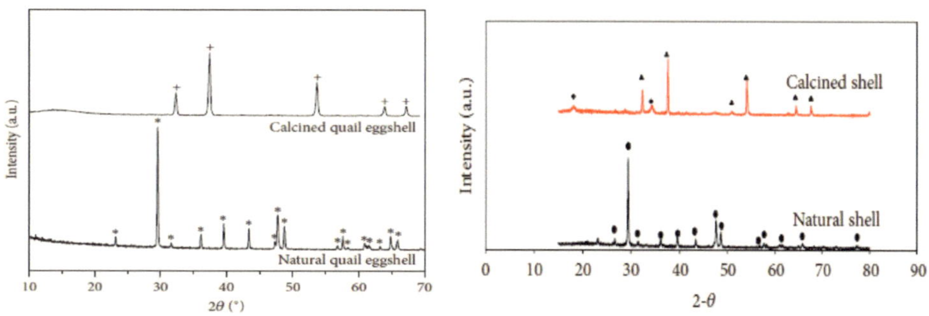

Figure 3.
XRD pattern of shells natural and calcined: (a) quail eggshell ($CaCO_3$, natural + CaO, calcined at 900°C) [23] and (b) oyster shell (symbols: •$CaCO_3$, ▲CaO, and ♦$Ca(OH)_2$) [21].*

and a small amount of $Ca(OH)_2$. Both the quail eggshell and the oyster shell share identical diffraction patterns for both the natural and calcined forms.

This suggests a similar mineralogical identity. After calcination (thermal treatment process), the diffraction lines attributed to rhombohedral phase for $CaCO_3$ disappeared, with new diffraction patterns arising around 2θ = 32.3°, 37.4°, 53.7°, 63.9°, and 67.3° assigned to cubic phase for lime (CaO) appeared (**Figure 3**). It is worthy to note that the quail eggshell exhibited a crystallite size of 315 nm ($CaCO_3$), while its calcined counterpart showed a size of 240 nm, CaO [23]. This crystallite size decrease can be ascribed to the exothermic natures of the calcination process. However, the lower intensity peaks for calcined eggshell and oyster shell could be related to the reduction in the crystallite size [21, 23]. Hence, the changes in the XRD pattern as a result of calcination are because of the release of carbon dioxide from the decomposition of $CaCO_3$ into CaO.

3. Methods of sorbent preparation

The associated complexity and high cost for the production of carbon dioxide capture adsorbent materials such as activated carbon or zeolite has shifted attention to exploiting and developing cheap and renewable materials such as eggshells and seashells biomaterials. **Figure 4** shows the procedure involved in the preparation of sorbent material from eggshells and seashells. The waste eggshells and seashells first undergo pre-treatment, which begins with acetic acid treatment with a concentration in the range of 1–10 molar to remove dirt, membrane layer, fibrous matters, proteins and other impurities as well as improve pore structure of the biomaterial [24]. Exposing the waste shells to acetic acid promotes the detachment of protein-collagen membrane depending on the extent, concentration and duration. At the end of this process, the sample is filtered and rinsed with distilled or deionised water. The separated eggshell or seashell is dried at 100–200°C for 5 h [16]. The dried biomaterials are crushed and then sieved into different particle size ranges depending on the application. The particles are calcined; the calcination process involves heat treatment to decompose the major component $CaCO_3$ into CaO. The temperature of calcination could range from 500 to 1000°C depending on the application. It has been reported that at 900°C, the $CaCO_3$ undergoes complete conversion into CaO [21]. The material produced after calcination is the sorbent material, which is placed in a desiccator to curtail the chances of coming in contact with humidity and carbon dioxide in the air.

In the pre-treatment phase, the reaction of acetic acid with $CaCO_3$ results in the formation of calcium acetate, which has a larger molar volume than $CaCO_3$ and CaO [25]. The acetic acid treatment helps to expand and improve particle pore structure. As a result of the expanded and enhanced pore network structure, improve performance is achieved over multiple carbonation-calcination reaction

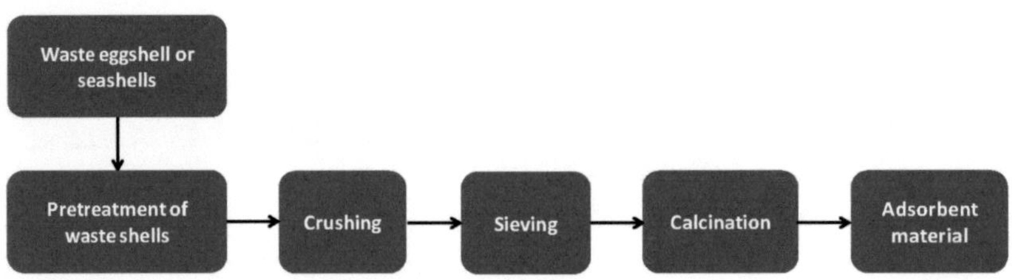

Figure 4.
Adsorbent material preparation procedure from eggshell and seashells.

(CCR) cycles [24, 26, 27]. Hence, the increased porosity within the microstructure of the synthesised CaO sorbent biomaterial from eggshell or seashells leads to increased reactivity over time.

4. Adsorbent performance

The continuing reliance on fossil fuels such as coal, natural gas and crude oil emits greenhouse gas (GHG) especially carbon dioxide (CO_2), a major contributor to global warming. The application of physical and chemical absorption using solvents such as selexol, rectisol, and mono-ethanol-amine (MEA) to remove carbon dioxide from flue gas streams is limited by low-temperature, cost and energy-intensive to regenerate [11]. Produced CaO sorbent material from eggshells or seashells through the method outlined in **Figure 4**, has proven a good candidate for carbon dioxide capture from flue gas stream of power plants. This is owing to their affinity to carbonate in the presence of CO_2; resulting in the formation of $CaCO_3$ which is regenerated back to CaO via calcinations while pure CO_2 is released for sequestration in the process as shown in **Figure 5**.

Unlike the adsorption process for CO_2 capture using activated carbon or zeolite adsorbent materials, eggshells and seashells biomaterials are low-cost and offer exclusive environmental and economic benefits. Additionally, eggshell or seashell-derived CaO sorbent are abundant, renewable, simple to prepare and also possesses excellent thermal stability. The mechanism of CO_2 capture by these biomaterials comprises of a series of carbonation-calcination reactions (CCR): calcium oxide (CaO) derived from eggshell or seashell reacts with CO_2 in the flue gas stream, leading to calcium carbonate ($CaCO_3$), which then undergoes calcination resulting in the release of a pure CO_2 stream for sequestration, and at the same time is regenerated into CaO as shown in **Figure 5** [24]. The pilot-scale demonstration of the concept has been reported for eggshell and oyster shell in the literature [24, 26–29]. The reactions are summarised as follows: carbonation ($CaO + CO_2 \rightarrow CaCO_3$) of the eggshell-derived CaO through reaction with CO_2 forms calcium carbonate ($CaCO_3$), while the calcination process ($CaCO_3 \rightarrow CaO + CO_2$), regenerates the CaO bio-composite material, and liberate pure stream of CO_2 for sequestration. Sacia et al. [27] investigated CaO sorbents derived from chicken eggshell for CO_2 from coal-fired power plants. In the work, they discovered that the pre-treatment of the eggshell with acetic acid enhanced

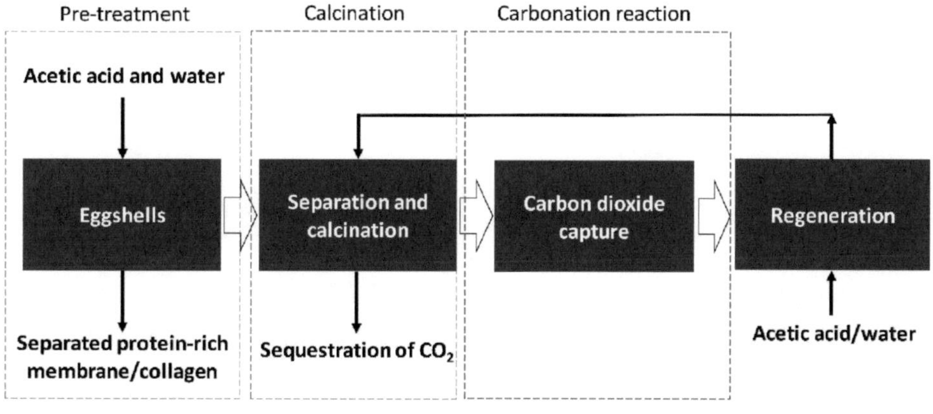

Figure 5.
Schematic of the eggshell or seashell carbonation – calcination processes for carbon dioxide capture.

and expanded the derived-sorbent material pore structure and surface area, which favoured CO_2 diffusion as mass transport is improved.

Figure 6 shows the effect of acetic acid concentration and treatment time on CO_2 capture over multiple cycles. It is clear that the acetic acid treated eggshell outperformed the untreated counterpart. On the other hand, derived CaO from eggshell treated with a low concentration of acetic acid exhibited better reactivity and CO_2 capture capacity than that treated with higher concentration. This can be attributed to the improved reactivity and porous surface structure within the biomaterials when treated with an optimised concentration of acetic acid [24, 26, 27].

Figure 6 also demonstrates that subjecting the eggshell or seashell to a higher strength acetic acid solution or for a longer treatment time could affect the pore structure, strength and stability of the derived CaO sorbent biomaterial. This is consistent with the result of the investigation reported by Sacia [17], on the use of eggshell for CO_2 capture. Hence, the observed decrease in the reactivity and CO_2 capture capacity under this condition. More also, the data shows that the derived sorbent from eggshell or seashell cannot be continuously regenerated over multiple cycles, as a result, fresh sorbent would be added as make-up during the process to sustain capture capacity (**Figure 6**). Depending on the acetic pre-treatment time, it has been reported that the CO_2 capture ranges from 70 to 80% in the first cycle, and gradually drop to about 40% in the fifth cycle [27].

Figure 7 shows simulated thermogravimetric analyser (TGA) results to prove CO_2 capture capacity of eggshell-derived sorbent using a typical flue gas stream (10% CO_2 for 60 min cycles at 700°C). The weight of the sample indicates reactivity, while the weight increase signifies carbonation due to CO_2 capture; the decrease represents the calcination process because of CO_2 liberation. It is clear that the CO_2 capture performance and reactivity gradually diminishes for multicycles over time.

The reactivity and CO_2 capture capacity of the eggshell or seashell derived CaO sorbent decline over time, so regeneration of sorbents in-situ is pivotal to maintaining CO_2 capture. The regeneration can be carried out using deionised water and acetic acid solutions [27]. The effect of regeneration of the eggshell derived CaO sorbent on CO_2 capture is shown in **Figure 8**. It is clear that regeneration with acetic acid is more effective than with water. Sacia [17] ascribed this observation to two factors. First, the use of acetic acid resulted in calcium acetate, which exhibited a higher molar volume than only $Ca(OH)_2$ formed when water is used. Also, the combination of water and acetic acid allows for a surface structure rearrangement due to the solubility of calcium acetate in water. It has been found that the use of

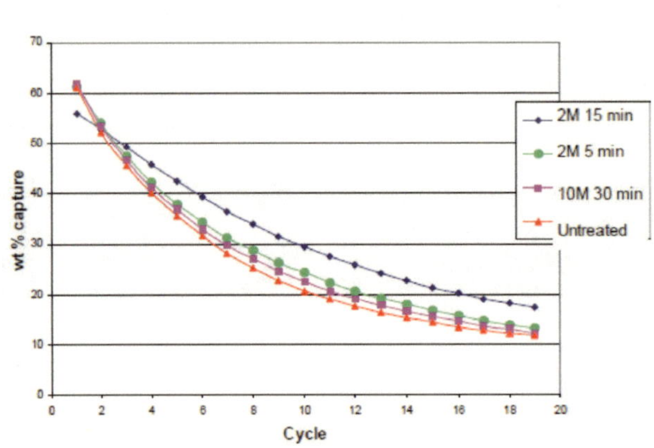

Figure 6.
Effect of acetic acid and treatment time on weight per cent CO_2 capture using chicken eggshell [24].

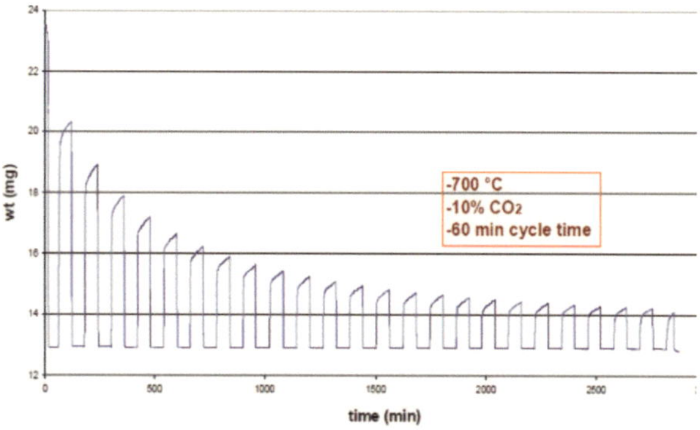

Figure 7.
Weight vs. time of eggshell-derived adsorbent for CO_2 capture using TGA [24].

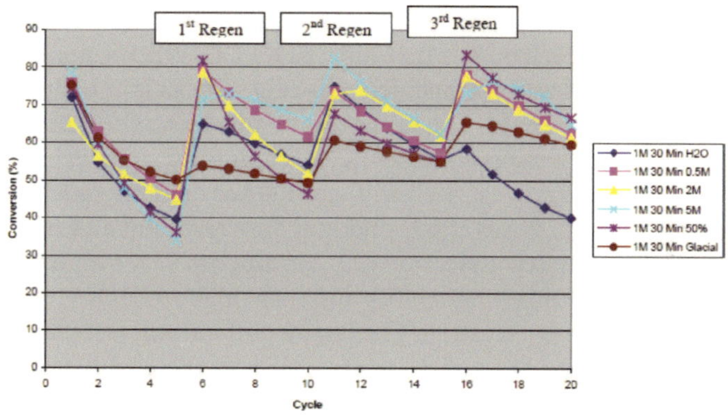

Figure 8.
Conversion vs. regenerations of eggshell derived CaO sorbent treated with a 1 M acetic acid for 30 min [17].

2 M acetic acid offers the best performance after multiple cycle regeneration in terms of reactivity and CO_2 capture [17, 24]. It can be observed that over three regenerations, all of the sorbent showed similar results trend.

In the investigation of Sacia et al. [27], it was found that regeneration restored the reactivity of the eggshell-derived CaO sorbent, and subsequently, CO_2 capture capacity in the range of 70–80% was achieved. The CO_2 capture capacity increased on average after successive regeneration, as can be seen in **Figure 8**. This suggests that periodic regeneration can effectively increase the reactivity of the spent eggshell or seashell-derived CaO sorbent. In another study by Banerjee et al. [30], it was reported that after four successive regenerations over multi-cycles usage, the carbon dioxide capture capacity of the eggshell-derived sorbent material decreased from 6824 mg CO_2/g to 1608 mg CO_2/g an average compared to the fresh material. This indicates that the eggshell-derived CaO sorbent biomaterial could hold about eight times its own weight of CO_2 from flue gas. Furthermore, Ma and Teng [31] investigated and reported the carbonation – calcination loop of CaO/$CaCO_3$ process for CO_2 capture using CaO derived sorbent from oyster shells. Though compared to reagent grade CaO from $CaCO_3$, the oyster shell derived CaO possess bigger crystallite size and lower specific surface area. It was reported that at 740°C carbonation temperature, the oyster shell-derived CaO sorbent in cyclic carbonation exhibited superior performance to the reagent-grade CaO obtained from $CaCO_3$. Therefore, utilising this waste biomaterial in CO_2 capture encourages the reuse of materials in the industries, which will reduce the risk, cost and energy associated with mining limestone and dolomite for $CaCO_3$ and CaO, and subsequently offers economic and environmental benefits. However, these benefits will be significant if the system is scaled-up to industrial standards.

5. Conclusion

There are large tonnes of eggshells and seashells discarded in landfill annually from poultry and food industries. Most of the seashells and eggshells are piled up on the seashore and thus would cause risks to water resources and public health. The applications of these biomaterials in construction such as concrete and cement production, catalyst manufacture, adsorbent for wastewater treatment, source of calcium in animal feed, manufacture of hydroxyapatite biomaterial, and additive in plastic manufacture has been explored extensively in the literature. These biomaterials contain about 96% calcium carbonate mineralogical component from which calcium oxide can be produced through thermal treatment. The carbonation – calcination loop of $CaO/CaCO_3$ process has been investigated for CO_2 capture potentials. Herein, the application of eggshell and seashell derived-CaO sorbent in the capture of carbon dioxide from flue gas is reviewed. The utilisation of this waste shell offers economic as well as environmental benefits because they are abundant, renewable and cheap. The CaO sorbent derived from eggshell and seashell has demonstrated the potential for carbon dioxide capture. It was also found that pre-treatment and regeneration provide means of restoring reactivity and CO_2 capture capacity over multicyclic usage. Although this ensured sustainability and sorbent recyclability, the performance decreases ten cycles after regeneration. The future outlook will be to improve the carbon dioxide capture capacity and thermal stability of these biomaterials over multicycles operations.

Author details

Abarasi Hart and Helen Onyeaka*
School of Chemical Engineering, University of Birmingham, Edgbaston, Birmingham, United Kingdom

*Address all correspondence to: onyeakah@bham.ac.uk

References

[1] COP21 Paris Agreement, European Commission, http://ec.europa.eu/clima/policies/international/negotiations/paris/index_en.htm (accessed 10/07/2020).

[2] IEA (2018) The Future of Petrochemicals: towards more sustainable plastics and fertilisers (executive summary). 2018, International Energy Agency

[3] Bui M., et al. (2018) Carbon capture and storage (CCS): the way forward. Energy & Environmental Science, 11 (5): 1062-1176.

[4] Hart A., Wood J. (2018) In Situ Catalytic Upgrading of Heavy Crude with CAPRI: Influence of Hydrogen on Catalyst Pore Plugging and Deactivation due to Coke. Energies, 11, 636.

[5] Khan S.A.R., Zhang Y., Kumar A., Zavadskas E., Streimikiene D. (2020) Measuring the impact of renewable energy, public health expenditure, logistics, and environmental performance on sustainable economic growth. Sustainable Development, 28: 833-843.

[6] Khan S.A.R., Sharif A., Golpîra H., Kumar A. (2019) A green ideology in Asian emerging economies: From environmental policy and sustainable development. Sustainable Development, 27: 1063-1075.

[7] Zhang Z., Pan S.-Y., Hao Li H., Cai J., Olabi A.G., Anthony E.J., Manovic V. (2020) Recent advances in carbon dioxide utilization. Renewable and Sustainable Energy Reviews, 125: 109799.

[8] Khan S.A.R., Zhang Y., Belhadi A., Mardani A. (2020) Investigating the effects of renewable energy on international trade and environmental quality. Journal of Environmental Management, 272: 111089.

[9] Khan S.A.R., Zhang Y., Sharif A., Golpîra H. (2020) Determinants of economic growth and environmental sustainability in South Asian Association for Regional Cooperation: evidence from panel ARDL, Environmental Science and Pollution Research, https://doi.org/10.1007/s11356-020-10410-1.

[10] Tcvetkov P., Cherepovitsyn A., Fedoseev S. (2019) Public perception of carbon capture and storage: A state-of-the-art overview, Heliyon, 5: e02845.

[11] Salaudeen A.S., Acharya B., Heidari M., Al-Salem M.S., Dutta A. (2020) Hydrogen-Rich Gas Stream from Steam Gasification of Biomass: Eggshell as a CO_2 Sorbent. https://dx.doi.org/10.1021/acs.energyfuels.9b03719

[12] Caldwell J.S. (2014) Experimental and computational evaluation of activated carbons for carbon dioxide capture from high pressure gas mixture. PhD thesis, University of Birmingham.

[13] Samanta A., Zhao A., Shimizu G.K.H., Sarkar P., Gupta R. (2011) Post-Combustion CO_2 Capture Using Solid Sorbents: A Review. Industrial & Engineering Chemistry Research 51: 1438-1463.

[14] Krutka H., Sjostrom S., Starns T., Dillon M., Silverman R. (2013) Post-Combustion CO2 Capture Using Solid Sorbents: 1 MW Pilot Evaluation. Energy Procedia, 37: 73 – 88.

[15] Blamey J., Anthony E.J., Wang J., Fennell P.S. (2010) The calcium looping cycle for large-scale CO_2 capture. Progress in Energy and Combustion Science, 36: 260-279.

[16] Hart A. (2020) Mini-review of waste shell-derived materials' applications. Waste Management & Research, 38 (5): 514 – 527.

[17] Sacia R.E. (2009) Synthesis and regeneration of enhanced eggshell sorbents for clean coal application. BEng thesis, Ohio State University.

[18] Trzepizur P.K. (2017) Waste Ca (Eggshells) natural materials for CO_2 capture. MSc thesis, Tecnico Lisboa.

[19] Arias B., Alonso M., Abanades C. (2017) CO_2 Capture by Calcium Looping at Relevant Conditions for Cement Plants: Experimental Testing in a 30 kWth Pilot Plant. Industrial & Engineering Chemistry Research, 56: 2634-2640.

[20] Laskar I.B., Rajkumari K., Gupta R., Chatterjee S., Paul B., Rokhum L. (2018) Waste snail shell derived heterogeneous catalyst for biodiesel production by the transesterification of soybean oil. RSC Advances, 8: 20131-20142.

[21] Buasri A., Rattanapan T., Boonrin C., Wechayan C., Loryuenyong V. (2015) Oyster and *Pyramidella* Shells as Heterogeneous Catalysts for the Microwave-Assisted Biodiesel Production from *Jatropha curcas* Oil. Journal of Chemistry, 578625.

[22] Tan Y.H., Abdullah M.O., Nolasco-Hipolito C., Taufiq-Yap Y.H. (2015) Waste ostrich- and chicken-eggshells as heterogeneous base catalyst for biodiesel production from used cooking oil: Catalyst characterization and biodiesel yield performance. Applied Energy, 160: 58-70.

[23] Correia M. L., Cecilia A. J., Rodríguez-Castellón E., Cavalcante Jr L.C., Vieira S. R. (2017) Relevance of the Physicochemical Properties of Calcined Quail Eggshell (CaO) as a Catalyst for Biodiesel Production. Journal of Chemistry, 5679512. https://doi.org/10.1155/2017/5679512.

[24] Vonder-Haar T. A. (2007) Engineering Eggshells for Carbon Dioxide Capture, Hydrogen Production, and as a Collagen Source. B.S. Dissertation, Ohio State University, Columbus, OH.

[25] Sasaoka E., Uddin M. A., Nojima S. (1997) Novel Preparation Method of Macroporous Lime from Limestone for High-Temperature Desulfurization. Industrial & Engineering Chemistry Research, 36: 3639-3646.

[26] Iyer M.V., Sparks A., Vonder-Haar T., Fan L.-S. (2005) High Temperature CO_2 Capture using Waste Oyster Shells. (Bunri Gijutsu) Soc of Sep. Proc. Eng. Jpn, 35 (4): 235-245.

[27] Sacia R.E., Ramkumar S., Phalak N., Fan L.-S. (2013) Synthesis and Regeneration of Sustainable CaO Sorbents from Chicken Eggshells for Enhanced Carbon Dioxide Capture. ACS Sustainable Chem. Eng., 1: 903-909.

[28] Wang W., Ramkumar S., Li S., Wong D., Iyer M., Sakadjian B. B., Statnick R. M., Fan L.-S. (2010) Subpilot Demonstration of the Carbonation-Calcination Reaction (CCR) Process: High-Temperature CO_2 and Sulfur Capture from Coal-Fired Power Plants. Industrial & Engineering Chemistry Research, 49: 5094-5101.

[29] Charitos A., Rodriguez N., Hawthorne C., Alonso M., Zieba M., Arias B., Kopanakis G., Scheffknecht G., Abanades J. C. (2011) Experimental Validation of the Calcium Looping CO_2 Capture Process with Two Circulation Fluidized Bed Carbonator Reactors. Industrial & Engineering Chemistry Research, 50: 9685-9695.

[30] Banerjee A., Panda S., Sidhantha M., Chakrabarti S.,

Chaudhuri B., Bhattacharjee S. (2010) Utilisation of eggshell membrane as an adsorbent for carbon dioxide. Int. J. Global Warming, 2 (3): 252-261.

[31] Ma K.-W. and Teng H. (2010) CaO Powders from Oyster Shells for Efficient CO_2 Capture in Multiple Carbonation Cycles. J. Am. Ceram. Soc., 93 (1): 221-227.